111 Questions and Answers in Packaging Technology

by Sola Somade and Tunji Adegboye

iUniverse, Inc.
New York   Bloomington

111 Questions and Answers in Packaging Technology

iUniverse books may be ordered through booksellers or by contacting:

iUniverse
1663 Liberty Drive
Bloomington, IN 47403
www.iuniverse.com
1-800-Authors (1-800-288-4677)

ISBN: 978-0-595-52684-0 (pbk)
ISBN: 978-0-595-51568-4 (cloth)
ISBN: 978-0-595-62738-7 (ebk)

Printed in the United States of America

iUniverse rev. date: 05/29/2009

Preface

Packaging is an exciting, dynamic, and challenging profession. It is an interdisciplinary subject that requires a reasonable knowledge of chemistry, physics, engineering, mathematics, marketing, economics, graphic design, and industrial processes among other subjects. The world now realizes the importance of packaging, not only as a factor in food sustainability, but also as the backbone of international trade in goods and services. Therefore, more knowledgeable people need to be involved in the packaging profession to help protect the consumers, our natural resources and the environment through the development and design of packages that are environmentally friendly and still provide the much-needed protection for our goods.

111 Questions and Answers in Packaging Technology is not a textbook. It is written with the primary aim of contributing to the success of those who are aspiring to become packaging professionals through formal and informal training. Trainers should find it handy as a reference book. Above all the book should serve as a useful handbook and a reference to anyone involved in packaging one way or another. For example, the brand/marketing manager, the packaging buyer, and the quality assurance manager should find the chapters on economics and quality assurance and specifications of tremendous interest.

The questions and answers in this book are based on our thirty years experience in handling packaging matters at both Unilever Nigeria Plc and Cadbury Nigeria Plc. Some of the questions are drawn from past question papers and lecture notes of the Institute of Packaging, UK, while many are drawn from our hands-on experience as packaging developers, quality managers, and packaging buyers at our previous places of work.

Some chapters have been specifically included in the book to help beginners as well as old practitioners to demystify some key areas of packaging that continue to cause anxiety. One such area is the chapter on 'Packaging Economics and Related Calculations'. A good under-standing of basic calculations will assist practitioners in making sound economic decisions on packaging material choice. Another key area is packaging specification writing, which we have made part of chapter ten. While we are not unmindful of the fact that no one packaging specification format has been universally adopted, we have no doubt in our minds that the format we have presented in this book will assist individuals to develop their own formats with the appropriate language that will be understood by their clients and other stakeholders. Other chapters have highlighted issues that are relevant to each of the major packaging materials known to modern civilization, such as metal, glass, paper, and plastics.

In writing this book, we see the entire English-speaking packaging world as our constituency. Therefore, the British system of spelling and the metric system of units have been adhered to, while the US dollar as the monetary currency has been adopted in all economic calculations in the book.

In preparing this book, several individuals/organizations played vital roles in the professional lives of the authors. We appreciate the op-portunities given to us by our previous employers to serve them in various areas of packaging. The assistance and encouragement of senior colleagues at both Unilever and Cadbury early in our packaging careers is well acknowledged. The late Mr D.W. Shorten, an IOP-appointed tutor for Tunji, is remembered and appreciated for his assistance, en-couragement and leadership quality.

We hope our readers find this book of much assistance.

Sola Somade
Tunji Adegboye
May 2008

Contents

Chapter 1 Basic Facts about Packaging

Question 1 – Give a definition of packaging and state four main functions of packaging. How are these functions exemplified in the packaging of a bar of chocolate wrapped in plain aluminium foil, twelve bars packed in a printed display box and six display boxes packed in a corrugated outer case?

Packaging is a coordinated system of preparing goods for storage, transport, distribution, retail, and use.

The four main functions of packaging are

(i) The containment function, which requires that the packaging material must adequately contain the product.

(ii) The protection/preservation function, which requires that the packaging material must protect/preserve the product it contains. This function is widely believed to be the primary function of packaging.

(iii) The transportation function, which is more or less an extension of the protection function because the packaging material must ensure the safe delivery of the product, no matter the transportation mode and rigour.

(iv) The information function, which provides all the relevant details about the product. Without this the consumer may have difficulty deciding whether to buy a product.

Chocolate is a heat-sensitive product and also highly susceptible to physical and bacteriological damage. Packaging must of necessity prevent all likely damages or deterioration during the shelf life of the product.

The aluminium foil wrapper, apart from containing the bar, is capable of preventing deterioration by moisture and oxygen ingress as well as any bacteriological deterioration. The printed display box offers some protection against physical damage. With proper branding and labelling, the box also informs the consumer, who with the help of such information decides whether to buy the product.

The corrugated shipping container makes product transportation and distribution possible with minimum physical damage to the bars.

Question 2 – What sets of facts have to be considered before an effective package can be devised for local distribution of a product? What facts about an apple would you consider relevant in the choice of packaging?

Before an effective package can be devised for local distribution, one must know

(i) Facts about the product,

(ii) Facts about the method of distribution, and

(iii) Marketing considerations.

The facts that are relevant in choosing a packaging for an apple for local distribution are facts about the apple itself. It is assumed that the apple is a salesman in its own right. The apple can suffer mechanical damage if it is not properly packaged. Mechanical damage can also occur before and/or after it has been packaged if it is not carefully handled. The resulting bruises/wounds will make it deteriorate quickly. It should be noted that an apple packed in an airtight enclosure will quickly suffocate. In addition, if the airtight enclosure is hot, ripening will be accelerated, which will lead to early product damage or deterioration.

Finally facts about the method of distribution will help to ensure a suitable design of a package for the apple, particularly as far as mechanical damage is concerned.

Question 3 – 'Packaging should always be considered while the product is still in the "blue-print" stage and is an integral part of the production process'. Write a short essay showing the applicability of this statement in respect of the pharmaceutical industry.

The above statement is very relevant to product development in any industry. But it is more so in the pharmaceutical industry where most of the products being handled end up being consumed by sick people. The need for product/packaging compatibility becomes more compelling because of the danger any error of omission or commission could pose to the health of the consumer. In order to prevent such ugly incidence, packaging has to be considered at the same time as the product is being developed. This must be so because a lot would have been lost if product/packaging incompatibility, or any packaging inadequacy, is established at the time the product is already in distribution. Another factor is the ability of the packaging to withstand warehouse, transport, and distribution hazards.

The whole idea behind early packaging consideration is to ensure that faults, if any, are detected and corrected before it is too late. It will be a costly experience if a product is already launched before its weaknesses are discovered. Apart from the quantifiable losses incurred, the loss of goodwill to the company is immeasurable and may even reflect negatively on the company's other products already on the market.
Also, since packaging constitutes a reasonable proportion of the overall product cost, a cost-effective package could mean the difference between success and failure of a product launch.

In summary, packaging must

(i) Be compatible with the product

(ii) Protect and preserve the product

(iii) Withstand warehousing and distribution hazards

(iv) Be cost-effective and make the product competitive

(v) Be presentable as well as informative to the consumer

Question 4 – *Compatibility between product and packaging is an important packaging principle. Give one example where the product might be affected by the package and another example where the product might affect the package.*

One example where a product might be affected by the package is in respect of orange squash where it has been observed that the squash does not last long when packed in polyethylene (PE) bottles as when packed in glass bottles. This is because PE bottles are permeable to oxygen and water vapour, and this leads to early product deterioration and rancid taste.

One example of how the product may affect the package is where unwrapped low Total Fatty Matter (TFM) soap is packed in an unlined corrugated case. The soap loses moisture to the inner walls of the case, thus dampening and weakening it. The moistened board becomes mechanically weak and is a potential medium for mould growth, particularly in a tropical environment. Any mould development will in turn adversely affect the quality of the soap itself. The phenomenon therefore is a double-edge sword.

Question 5 – *What material properties would you need to take into account when deciding on a printing process for the following?*

    (a) Low density polyethylene film

    (b) Collapsible metal tubes

    (c) Reel-fed, heat-sealed paper labels

Indicate the printing processes, which are compatible with, and commercially viable for, each material.

(a) Low density polyethylene film – Material properties that need to be taken into account when deciding on a printing process are porosity or smoothness of the surface of the film; polarity of the film, if any; whether the material is to be reel- or sheet-fed; and the quantity to be printed.

The flexographic printing process is compatible with this material. Heat treatment is required before printing. LDPE has low soft point, hence no higher temperature drying of inks.

(b) Collapsible metal tubes – The surface quality of the tube must be taken into account. For instance, prior to printing, surface pretreatment may be necessary for the printing ink to stick, that is, the enamel quality will be an important factor.

Offset lithography will be a suitable printing process. This is because the tube surface is non-absorbent, flexing in use, and can stand high temperature drying (stoving).

(c) Reel-fed, heat-sealed paper labels – The smoothness of the paper, quality of print required, and the quantity of labels to be printed will all be important factors. Also the even distribution of the coating material will be essential for hitch-free heat sealing. The number of printing colours required is also a key factor.

The gravure printing process will be suitable for this job, particularly if the length of run is considerable. To avoid set-off, quick-drying inks are necessary. Inks must also be heat resistant.

Question 6 – Outline the essential requirements in marking and identifying the following:

(a) A unit pack for processed foods

(b) A shipping container for the export of heavy machinery

(c) Packaging for the defence services

(a) A unit pack of processed food – Essential requirements are that the marks should indicate

(i)   What the product is, that is, name of product or an already established brand name.

(ii) What is in it, that is, a declaration of all the ingredients in descending order by weight with a heading, 'Ingredients'.

(iii) The name and address of the packer, a company or person the purchaser could contact for information or to complain about the product, if necessary.

(iv) Anything special about the product or any other information that may be of relevance to the purchaser, such could be marked on the pack provided such claims comply with the 'Labeling of Food Regulation Act'. Examples are nutritional claims, declared weight, manufacture date and the best before or expiry date.

(b) A shipping container for export of heavy machinery – Marks on a shipping container for export should comply with the International Organisation for Standardisation. The idea is to ensure that language barriers are overcome. Hence the adoption of seven pictorial markings or symbols that people the world over, including those who cannot read, can understand. Part of the marks normally carries the name and address of the shipper and the consignee and what the package contains. Generally the pictorial marks give handling directives, including opening instructions.

(c) Packaging for the defence dervices – Here the sales appeal is less important. There are two essential requirements that must be satisfied:

(i) The identifying marks must be long lasting under all possible climatic conditions for a minimum of five years.

(ii) The identifying marks must be clearly seen and be unambiguous.

Question 7 – What are the different types of paper labels? Which label would one use for each of the following products and why?

(a) A frying pan

(b) A jar of coffee

(c) A packet of biscuits

(d) A ream of printing paper wrapped in bitumen kraft union

(e) A supermarket pricing label

Four main types of paper label exist:

(i) Plain paper label – Here the glue is applied manually or automatically to the plain label before the latter is applied to the object to be labeled.

(ii) Pre-gummed paper labels – Here the paper is pre-coated with dextrin or gum Arabic, which is normally activated by the addition of water.

(iii) Thermoplastic paper labels – These could also be called heat-sealable labels. Here the paper is pre-coated with synthetic thermoplastic resin, which is activated by heat application.

(iv) Pressure-sensitive paper labels, or the so-called self-adhesive labels – The paper is pre-coated with a permanently tacky adhesive and is applied to the product with pressure.

(a) A frying pan – Pre-gummed or pressure-sensitive paper labels will be suitable because no smearing of the surface will occur as could happen if a plain paper label is employed.

(b) A jar of coffee – All the paper labels listed above, except the thermoplastic one, will be suitable.

(c) End labels for a packet of biscuits – Pre-gummed or pressure-sensitive paper labels will be suitable. Using plain paper label could result in a poor labeling job, hence poor product presentation. The high temperature employed in thermoplastic labels could be too much for biscuit packets.

(d) A ream of printing paper wrapped in bitumen kraft union – Pressure-sensitive paper labels with adequate adhesive strength will be suitable. This will prevent wetting of the wrapping kraft and also ensure proper bonding.

(e) A supermarket pricing label – Pressure-sensitive labels with reasonable adhesion will be suitable. This is because it will be possible to remove or adjust the labels on the articles with little or no difficulty.

Question 8 – *The pack has been described as 'the silent salesman' in retail markets. Discuss this in not more than 500 words.*

Until recently, particularly in the developing nations, packaging has been considered a minor element in the marketing mix. The traditional objectives of packaging have been product protection and consumer convenience. Recent developments in retail practice have conferred a new important status on packaging – that of sales appeal. The sales appeal aspect of packaging is so important in modern retail selling for so many reasons that a lot of attention has to be devoted to the surface design of packs. This is necessary because in many cases it is the surface design of a pack that first appeals to consumers rather than the product itself, which of course is alien to the first-time buyers. In such situations the package surface design makes a lot of difference between selling and not selling.

In addition to sales appeal, the pack has to carry some identifying statements voluntarily or involuntarily, without which the pack does not make any sense to the prospective buyers. Such information or identifying statements include

(i) The brand name of the product and its producer/packer.

(ii) A declaration of product ingredients, particularly for edible products where preservatives, colourants, and so forth must also be declared for public health and safety reasons.

(iii) Nutritional claims on vitamins, minerals and other essential ingredients on the basis of Recommended Daily Allowance (RDA).

(iv) Directions for use – this is particularly important in medicinal and poisonous products.

(v) Package average or minimum packed weight or filled volume are required to be declared on the pack where the Weights and Measures Acts are in force.

(vi) Other pieces of information the package carries are the manufacturing date, the batch number for traceability in the event of quality problems, and the best before date to guide the consumer.

(vii) In some sophisticated economies, prices are now printed on the individual packs for the convenience of shoppers at supermarkets and other self-service stores. As an alternative, a bar code, which carries a lot of information on the product, is incorporated into the package design. Scanning machines for decoding such information are now part of the facilities of most supermarkets worldwide.

All the above points about pack design have helped to reduce the burden of consumers considerably. For instance, consumers no longer need to call the attention of the storekeepers, as it used to be, before deciding whether or not to buy a product because all the relevant information required to make a decision are already on the pack. The presence of a price tag or label has also helped both the storekeeper and the buyer to overcome the problem of spending valuable time haggling over the price of an item.

Considering the roles the pack design/labeling or identification plays in modern self-service retail outlets—informing, advising, and advertising—packaging is truly playing the role of the silent salesman.

Question 9 – How is a folding paperboard carton (retail pack) made? What effect does each stage in the manufacturing process have on the performance of the carton during filling and in subsequent use?

Once the type of material and the blank size for the carton are chosen, the making of carton passes through the following stages: printing, cutting and creasing, stripping and finishing (folding and gluing) operations.

In choosing a material, the product characteristics and the packing arrangement must be taken into consideration. Grammage and the caliper of the board are selected at this point, and if printing is to be carried out on a litho machine, the size of the board sheet will have to be specified. This is usually dictated by both the machine deckle, carton size, and the number of ups (cartons) per board sheet.

Printing – Many printing methods exist for this kind of job. However whichever printing process is adopted, the resulting print must have the required resistance to rubbing during the filling and sealing operations on the packing line. It must also be resistant to the rubbing it will receive during transportation/handling. The choice of ink drying and handling must be such that ink set-off, smudging, and other print defects are minimized.

Cutting/Creasing – The purpose of these operations is to prepare the printed carton sheet to be cut into layout with the resulting layout detached from the sheet or web with creases correctly located so that the carton may be readily glued as required for subsequent symmetric erection on the packing machine. The die-cutting formes could be a block die, made from metal or plastics, and is usually cut to the size of the carton panels with the rules clamped between the blocks. A forme has rows of cutters arranged to correspond to the printed images on the sheet. The crease is made by the rounded-edged rule pressing the board into a groove in the make-ready or counter. These creases serve two purposes. First, they fix the lines at which the board will fold easily, and second, they prevent the board from cracking at the bends.

Stripping – The result of the cutting and creasing operations on a sheet of board is to produce a number of cartons held together by an intact front edge and small bridges at various points on the sheet, so that it can still be handled as a sheet and does not fall completely apart into its component cartons. The stripping operation is to remove the material which holds the sheet together. Even though part of the stripping is done as the sheet passes through the cutting and creasing press, this aspect of carton making is usually completed by hand. Essentially the operator uses a rubber-headed hammer and knocks the waste material away from the cartons, which are torn out of the sheet in blocks. This takes us to the next operation. On modern gravure printing presses, however, stripping is done automatically online.

Finishing – This is essentially a folding, gluing, and packing operation. For cartons meant for erection on automatic packing machines, this is an important operation. When a collapsed tube is formed on the side seam gluer, two diagonally opposed creases are folded 180°. The majority of gluers have a pre-folding operation, which momentarily fold the other two main creases to between 90° and 180°, and then unfolds them. By this means, the subsequent erection of the carton on the packing line is made easier. This pre-folding operation is referred to as pre-breaking the creases. After the gluing operation, the cartons are packed in well-arranged form in cases or boxes ready for dispatch to the user.

The effects each stage can have on the carton performance are as follows:

(i) Printing – Ink set-off or instability under certain liquid/wet media or direct sunlight may result if the printing ink is not stable. Poor post-printing drying may produce the same effect in addition to smudging.

(ii) Cutting/Creasing – Cutting must be carried out with a very sharp and straight cutter to avoid incomplete flap cuts or asymmetric cutting, both of which may lead to problems on the packing line. Creasing must be deep enough and straight so that cartons erect well and flaps do not spring back after folding and sealing on fast packing machines.

(iii) Stripping – No scraps should be left attached to the cartons since these also create problems on the packing lines. Also this is unhygienic and unacceptable on food packing machines.

(iv) Finishing – As previously stated, pre-breaking the creases does aid easy erection. Over- and under-application of glue on the side seam should be avoided as either leads to undesirable consequences such as machine jamming and attendant downtime on packing lines. Packing of cartons (preferably side-on) should be such as to avoid crushing or damaging the flaps.

Question 10 – What are the characteristics and limitations of composite containers? What materials would you choose for a composite container for the following?

(a) Dried milk

(b) Lubricating oil

Give reasons for your choice in both instances.

A composite container is made from more than one constituent material, generally consisting of a paperboard body with metal or plastic ends. For some functional reasons, the insides of the containers may be lined with another material. In short, a composite container is a marriage of two or more dissimilar materials to achieve a high functional need.

Two types of composite containers are available as defined by their method of production: spirally wound and convolutedly (straight) wound. While the spirally wound type is suitable only for cylindrical containers, in contrast, the convolutedly wound type can produce a wide variety of shapes, such as square, triangular, rectangular, and oval.

The body materials for both types are usually the chipboard or kraft paper with the former giving a more rigid body. One limitation of the containers is that the body materials offer virtually no protection against gas vapour or moisture ingress into the contents. Therefore for products that are sensitive to these environmental factors, special lining materials that are good barriers are essential. Another limitation is the inability of the unlined body materials to contain liquid contents, which easily permeate the porous body materials and therefore render it and the contents useless. Unfortunately, all the suitable lining materials are not yet extensively used due to their high cost.

(a) For dried milk, kraft paper lined with polyethylene-coated aluminium foil is suitable. This is because the product obviously requires an excellent water vapour and gas barrier to avoid deterioration due to absorption of moisture, gas, and oxygen.

(b) For lubricating oil, pure vegetable parchment used as a liner to chipboard is suitable. While the chipboard provides the strength and rigidity, the parchment offers good resistance to the oil so that the porous chipboard does not absorb the oil.

Question 11 – *What are moulded pulp containers? How are they manufactured and what are their relative merits when compared with thermoformed plastics containers?*

Moulded pulp containers are articles or containers moulded from a mixture of water and any type of fibrous material capable of being treated by the normal paper making processes.

The forming process used in the production of moulded pulp containers is, in many respects, similar to paper making but with the important difference that, whereas paper is formed on a traveling wire screen, moulded pulp containers are made in a mould fitted with a screen formed to the shape and profile of the container to be produced. There are two distinct moulding processes employed for the production of pulp containers: pressure injection and suction injection.

In the pressure injection process, the mouldings are produced on semi-automatic machines. Each machine is fitted with a mould normally containing six sections, of which five are movable. A measured amount of pulp and water mixture is admitted into the mould, and the article is formed by blowing in air at a pressure of approximately 4 kg/cm^2 and at a temperature of 480°C. The moulded articles leave the mould containing 45–50 per cent moisture and are subjected to an after-drying operation, which also sterilizes the mouldings. The mouldings produced by this method are usually strong, rigid, and light.

By contrast, in the suction process, the pulp mixture is deposited and formed to the shape of the mould by the application of a partial vacuum to one side of the mould screen. The formed article, as it leaves the mould, contains a much higher percentage of moisture, generally 85 per cent. The residual moisture has to be removed by an after-drying operation. The suction process is normally operated by fully automatic moulding machines, which require a minimum of two forming moulds and one transfer mould.

Some relative merits of moulded pulp containers compared to thermoformed plastics containers are as follows:

(i) In making moulded pulp containers, no intermediate finished or semi-finished product is involved whereas plastic sheets or resins need to be prepared prior to carrying out thermoforming operation.

(ii) Temperature control during making operations is not a sensitive factor in moulded pulp containers whereas temperature must be kept within a narrow range in thermoforming operations.

(iii) Moulded pulp containers are not prone to stress cracking whereas thermoformed containers are.

(iv) Moulded pulp containers are generally cheaper than thermoformed containers because recycled paper/pulp are in common use for moulded pulp containers whereas recycled plastics are hardly accepted/used in thermoforming operations.

Question 12 – What is a pressurised dispenser (aerosol) and how does it work? Describe briefly two methods by which pressurized dispensers can be filled, with the advantages and disadvantages of each.

Aerosol is an integral ready-to-use package incorporating a valve and a product that is dispensed by pre-stored pressure in a controlled manner when the valve is operated or activated. An aerosol package consists essentially of four parts: the container, the valve, the cap, and the propellant.

The container, made of aluminium or tinplate and constructed in a way similar to metal cans, holds both the product and the propellant (vapour or liquid).

The valve is a mechanical device, the operation of which permits the discharge of the product from the aerosol in a pre-determined manner.

The propellant is a material, which provides the power to eject the contents.

The cap, a removable protective cover over the valve actuator, prevents accidental operation of the valve.

Depressing the button over the valve uncovers the hole and connects the interior of the container to the atmosphere. The pressurized material inside the container is thus forced out because of the high pressure differential between the container interior and the atmosphere.

Filling Methods

Cold filling – This involves lowering the temperature of both the product and the propellant to below the boiling point of the propellant before filling. There is no special advantage known for this method. However, this process has a disadvantage that high capital cost and a large amount of energy are required to bring both the can and the contents back to room temperature before testing, labeling, and packing operations can be carried out.

Pressure filling – This involves handling and filling into open containers at ambient temperatures. After the product has been placed in the container, the air is removed. This is done either by replacing it with vapour propellant or by drawing a vacuum immediately prior to sealing the valve in place. The vapour purging method has the advantage of simplicity, but is more expensive to run since it involves the dropping of several grammes of propellant vapour into the can in order to displace the air before sealing. The exact amount of propellant vapour required varies with headspace volume and product.

The vacuum purging requires a more complicated crimping machine, and, unless this is a multi-head machine, it may limit the speed of production. Also it is more difficult to ensure that a satisfactory vacuum has been drawn.

Question 13 – What are the two main kinds of plastics, and what are their distinguishing characteristics? Give two examples of each kind of plastic.

The two main kinds of plastics are

(i) Thermoplastics

(ii) Thermosets

Thermoplastics have the properties of softening with the application of heat and hardening again when cooled. This can be done repeatedly. The long chain molecules of thermoplastics are held together by weak intermolecular Van der Waals forces. Heat energy can therefore cause these molecules to move relative to each other. Examples of thermoplastics are polyethylene (both low and high density types) and polypropylene.

Thermosets on the other hand become molten once on heat application and then harden irreversibly on withdrawal of heat. Further heating will only result in charring or burning of thermosets. Thermoset molecules, like those of thermoplastics, are long, but unlike thermoplastics, they are capable of further chemical reaction when heated. It is this reaction, which leads to a linking of the long chain molecules to each other to form a three-dimensional network. Examples of thermosets are phenol formaldehyde and urea formaldehyde.

Question 14 – Which plastics would you recommend for the following?

(a) A clear bottle, having good rigidity and resistance to the passage of oxygen

(b) A flexible bottle for dispensing powders

(c) The coating of paper, or a plastic film with high permeability to moisture vapour and gases, in order to give a material having a low permeability

(d) A bottle crate where resistance to long periods of stacking is required

(e) A film to be used for boil-in-the-bag packaging

Give reasons for your choice.

(a) Polyvinyl Chloride (PVC). This is because PVC is rigid, clear and has good gas barrier properties. PET can also be used.

(b) LDPE is quite flexible and so is an ideal container for dispensing powders.

(c) LDPE and polyvinylidene chloride (PVDC) both have the outstanding property of low permeability to water vapour and gases.

(d) Polypropylene copolymer and HDPE both have enough strength and resistance to impact to withstand long periods of stacking.

(e) Polyethylene Terephthalate (PET or polyesters) with polyethylene coating has high strength and a high softening point. HDPE can also be used.

Question 15 – Why is it sometimes necessary to incorporate additives in plastics? Give three examples of types of additives.

It is sometimes necessary to incorporate additives into plastics in order to enhance or improve their performance both during processing and use of the resulting mouldings or films. Examples of types of additives are

Plasticisers – These are often necessary to reduce the rigidity of some plastics (e.g. PVC) so that processing is easier at a lower temperature and so that permanent flexibility can be achieved. Examples of additives used as plasticisers for PVC are esters.

Stabilizers – These protect polymers against physical or chemical deterioration when subjected to atmospheric effects or to high temperatures during processing (e.g. PVC). Some of the stabilizers in use are organic and inorganic lead salts; organic derivatives of barium, cadmium, zinc and calcium; and organic derivatives of tetravalent tin.

Slip Additives – These are added to films in order to reduce the coefficient of friction between two film surfaces. Slip additives are usually long chain fatty-acid amides, such as polyolefins.

Question 16 – *What are the differences in heat-sealing behaviour between regenerated cellulose film and polyethylene film? How have the problems in sealing polyethylene film been overcome?*

While regenerated cellulose film can be heat sealed satisfactorily by heated metal jaws, polyethylene film cannot be heat sealed the same way because polyethylene sticks to hot metal surfaces. This sticking problem is solved by coating the metal jaws with a silicone rubber or with Teflon, poly-tetra-fluoro-ethylene (PTFE).

Another difference between regenerated cellulose film and polyethylene film is that while regenerated cellulose film has an infusible substrate and therefore only the coating melts on heating, polyethylene films are completely fusible, which results in loss of strength at the heat-seal area and leaves the seal unsupported while still hot and weak. This problem is overcome by the use of impulse sealers. Impulse sealers have jaws of light construction and have attached to them resistance wire or ribbon covered with PTFE. The resistance wire is heated electrically and because of their small heat capacity, the jaws heat up and cool quickly when switched on and off.

Question 17 – *How can the different properties of low density polyethylene and high density polyethylene be explained by reference to their molecular structures? Which of these two polymers would you use to make a bottle which had to be capable of withstanding steam sterilization, and why?*

The making of low density polyethylene is a high pressure, high temperature gas phase process. This results in a rather vigorous process that leads to a great deal of branching of the polymer chains as they are built up. This branching of the main chains prevents a close approach of the various chains to each other. Therefore the density of LDPE is lower than the Ziegler polyethylene, that is, HDPE.

High density polyethylene is made under low temperature/low pressure liquid phase conditions. Hence the polymer chains grow almost undisturbed and so form a substantially linear chain with very few (and much shorter) side branches. The closer packing of the molecular chains that characterize the HDPE means the material is denser and more rigid than the LDPE. Also, the close approach of the molecules of HDPE increases the attractive forces between them, and more heat energy is needed before this attraction can be sufficiently overcome for the individual molecules to flow relative to each other. This accounts for the fact that HDPE has a higher softening point than the LDPE.

HDPE is preferred to LDPE when there is need for steam sterilization because of its higher softening point.

Question 18 – Outline briefly the uses of waxes and bitumen in packaging.

Waxes and bitumen were successfully used for packaging in the ancient times, notably in closures for glass containers and for making paper and similar materials water resistant, as is the case in paper and board sizing. The need for their use arose from the fact that paper and board are very porous and, as such, are permeable to water vapour. Unfortunately, many products which are normally packed in paper or board nowadays are either hygroscopic (absorb moisture) or easily lose moisture; either way this is to the detriment of the product. Clearly there is strong need to reduce the amount of moisture that passes through the paper/board containers. Both waxes and bitumen, when used as laminating media between paper and board components are known to reduce the water vapour transmission rate (WVTR) considerably. This is because a wax/bitumen component of about 20 g/m^2 evenly applied to paper or board blocks the pores and therefore reduces the passage of moisture many folds. Hence moisture-sensitive products are assured of adequate protection for a fairly long period, particularly in hot and highly humid environments. An example of this is the use of wax-laminated chipboard for detergent powders in the tropical regions of the world.

Though bitumen is cheap, it has some disadvantages. Some of the disadvantages are its black colour and sticky nature. Bitumen gets transferred by the cutting knives from one place to another and rapidly makes equipment dirty. Another problem with bitumen is that when exposed to the sun's rays, it softens and is therefore unsatisfactory for outdoor conditions. Wax does not suffer any of the disadvantages of bitumen, but it is more expensive. Apart from its use in packaging, wax is also used in the formulation of hot-melt adhesive.

Question 19 – Outline the production of a blow-moulded polythene bottle. What variables in the container, which occur during manufacture, can influence its performance as a package, and in what way?

The basic techniques of blow moulding were derived from those developed by the glass industry. Air is forced under pressure into a sealed molten mass of plastic, which is surrounded by a cooled mould of the required shape. The air pressure causes the molten mass to expand. When it reaches the mould walls the plastic is cooled by contact. The mould is then opened and the bottle ejected.

There are now two main techniques in the plastic bottle blowing process: extrusion blow moulding and injection blow moulding.

Extrusion blow moulding – This process had a later start, but it now accounts for the largest proportion of blow-moulded objects produced today. It essentially consists of the extrusion of a pre-determined length of hollow plastic tube (parison). This is trapped by closure of a split die which seals both ends of the internal contours of the mould by compressed air introduced via a blowing pin. The moulding is again cooled by contact with the walls of the mould; the latter is opened and the bottle ejected.

Injection blow moulding – This is the process most closely resembling the blowing of glass bottles. Instead of extruding a parison, as in extrusion blow moulding, the plastic material from which the bottle is to be formed is moulded round a blowing stick in a converted injection moulding machine as a thick-walled tube called a parison. While still molten, the parison (or preform) and the blowing stick are transferred to a second or blowing mould, in which the final shape is obtained by passing compressed air down the blowing stick into the parison. This second/final blowing operation is similar to the extrusion blow moulding. The blown shape is cooled by the walls of the mould; the mould is opened and the bottle ejected. The cycle is then repeated.

Variables that can influence bottle performance are many. Bottle wall thickness and general plastic distribution are important. For instance when a length of plastic tubing of even wall thickness is blown into a bottle shape, the material is thinned more at the extremities of the mould than elsewhere. This factor is important because it is at these extreme points that extra strength is needed in the bottle. Also very important is the quality of the raw materials. For example, the resin must be free of foreign objects. Its density and melt flow index must be carefully chosen to have the required rigidity while at the same time ensuring no failure or stress cracking of the bottle during use.

Question 20 – What are two main methods or processes of producing plastic film? Briefly explain the two processes. What are the quality differences between the films produced by the two processes?

The two main processes for plastic film production are

  (i) Blown film process

  (ii) Cast film process

Both methods start from the extrusion of molten polymer; the difference only lies in the design of the die through which the molten polymer passes and then in the subsequent haul-off.
The plastic granules are fed through the hopper and are carried along the barrel by the rotation of the screw. As they progress, they are melted by contact with the heated barrel and by the generation of frictional heat.

Specifically in blow extrusion, which is the process used for the production of most film, the molten plastic from the extruder enters the die, where it is made to flow around a mandrel, emerging from a ring-shaped die opening in tubular form. The tube is blown into a bubble by air introduced via the mandrel. The air is trapped in the bubble by the die at one end and the pinch rolls at the other end. The size of the bubble and the thickness of the film are controlled by the extrusion speed, the haul-off speed, the die-ring width, and the air pressure within the bubble. The resultant collapsed tubular film, which is now sufficiently cooled, is known as lay flat film and can be used as such for bag or sack making. It can also be slit and made into flat film.

In slit-die extrusion (cast film making), the molten plastic is extruded through a slit-shaped die opening into a chilled metal roller, which rapidly cools the film. The slit-die extrusion method is the one usually employed for sheet manufacture.

Blown film method is very versatile and makes for the production of a wide range of film widths. The inflation and subsequent expansion of the bubble gives better balance of film orientation, resulting in improved impact strength. On the other hand, the rapid cooling in cast film making allows high production rates and better clarity. However, the uniaxial orientation of cast film results in lower toughness compared to blown film.

Question 21 – What is a laminate and why have laminates become the backbone of flexible packaging? What is the role of each layer in the Pet/Print/Foil/PE laminate, which has become very popular with oxygen and moisture-sensitive products?

A laminate is a flexible material made of two or more dissimilar components. Laminates came into the packaging scene due to the realization that no single-layer film can meet all the functional protection requirements needed by some products. The answer then lies in the use of compound materials, where the deficiency of one component is compensated by the strong point of the other. Since there are now many laminate structures serving many different needs, laminates have become the backbone of the flexible packaging. No matter how stringent or demanding the requirements, laminates now exist whose barrier properties are almost as good as those of glass and metal containers.

In the laminate structure Pet/Print/Foil/PE, the Pet layer provides a smooth surface for printing, a transparent surface through which the consumer can see the reverse printing, and a support to the weak jelly-like foil. The foil layer provides a good barrier against moisture, gases, etc. The PE layer acts as a sealant as well as providing some barrier against moisture and gases.

Question 22 – Describe the various methods of producing flexible laminates from film, foil, and paper and indicate the particular advantages of using such constructions.

The methods of producing flexible laminates from film, foil, and paper are

(i) Wet bonding

(ii) Dry bonding

(iii) Co-extrusion

(i) Wet bonding – This is a process where two or more webs are combined using an adhesive in the wet state. Usually at least one of the webs must be permeable to water or any other solvent used in the adhesive formulation so that drying can take place. Even in spite of this, the finished laminate must be run through a drying oven to speed up the drying process. The non-permeability of plastic films to solvents explains why they are not laminated by wet bonding method. This is a major disadvantage since plastic films are used extensively in many laminates.

(ii) Dry bonding – This is a process whereby a wet adhesive, hot melt adhesive, or one of a range of polymers is applied to one web, dried out, and then brought into contact with the other web. The combination is passed through nip rolls, which may be heated to speed up drying. Unlike wet bonding, this process is very suitable for the lamination of plastic films to other substrates. This is an advantage over the wet bonding method.

Major disadvantages of the dry bonding method are generally the need to maintain good tension control, accurate and even adhesive application, and accurate control of drying. Improper glue application and poor oven drying will all lead to de-lamination.

Another variety of dry bonding is extrusion lamination where a flat die extruder discharges a molten curtain of polyethylene, polypropylene, or some other thermoplastics into the nip between the two webs to be laminated. The heated adhesive is cooled by passing the laminated sheet over a specially designed 'combining and cooling' roll section. This cooling section replaces the drying section generally needed in wet adhesive bonding. Extrusion laminates are good barriers to water and water vapour and are usually tough and flexible.

(iii) Co-extrusion – This process consists of coupling two or more extruders to a single die head instead of producing the individual films at various extruding stations, reeling up and bringing the various films (in reels) together and unreeling them for lamination when needed. Using a single die head in place of two or more dies significantly lowers costs. Other advantages are a lower tendency towards de-lamination and a greater flexibility in obtaining a wide range of functional properties. Economic use of material is achieved because thinner films can be produced.

Disadvantages of co-extrusion process are

(a) Difficulty in utilizing the scrap generated during the extrusion process because it is difficult if not impossible to effect separation of films thus produced.

(b) Laminates with sandwich printing cannot be produced.

(c) Film (or laminate) produced by co-extrusion is only economical if the run is long.

Question 23 – What is shrink-wrapping? What are the advantages and disadvantages of shrink-wrapping vis-à-vis fibreboard cases as transit packs for twelve glass jars of jam?

Shrink-wrapping is a process whereby a loose hood of plastic film (e.g. polyethylene) is wrapped round a load of product packs and the pallet or tray (on which the packs are arranged) and then shrunk by heat application to hold tightly on the tray/pallet pack.

Advantages of shrink-wrapping over fibreboard cases: It is cheaper than any other shipping containers known, including fibreboard cases. It is suitable for a wide range of products or loads regardless of their shapes. The wrapping film is transparent and therefore makes inspection easy without opening the pallets/trays. It also offers good weather protection and keeps goods free of dust.

Disadvantages of shrink-wrapping rather than using fibreboard cases: It does not provide good physical protection because it has no cushioning effect. Therefore with products like glass jars, breakage is very likely if handling is not carefully done. With fibreboard cases, the use of internal dividers can substantially reduce product damage. With time, the shrunk film can suffer some relaxation resulting in looseness of shrink-wrapped packages.

Question 24 – Discuss the various types of closure possible with corrugated fibreboard outer containers, outlining the advantages and disadvantages of each method.

There are three main ways of providing effective closure for corrugated fibreboard cases. The flaps, particularly the outer ones, can be secured in place by

(a) The application of adhesive tapes

(b) Stitching or stapling

(c) Glue application

(a) The adhesive tape can be the self-adhesive type with a plastic film base or kraft paper base requiring water to activate the adhesive coating prior to sticking the tape to the case. These adhesive tapes are usually applied by machines (case sealers) along the lines where the case flaps meet.

Advantages of tapes are

(1) They provide effective closure that does not permit flaps to flip open.

(2) No gaps are left between flaps; therefore, access to the case contents is not easy. Pilferage is thus prevented.

(3) Tapes can be branded by pre-printing to complement the case design and enhance the corporate image of the company and its product.

(4) Tapes can also act as tamper evidence devices.

The only known disadvantage is that self-adhesive tapes are generally more expensive than other closure methods.

(b) Stitches can be applied to case flaps manually or by machines similar to those used for stitching the manufacturer's joints.

Stitching has only one known advantage: it provides a good and effective closure.

Disadvantages are

(1) Delicate products could get damaged if they have direct contact with the stitches, more so if the latter are not blunt enough.

(2) Opening the cases requires more effort with stitches than with tape or glue.

(3) Use of wire stitches for cases containing food items are generally not accepted on account of Hazard Analysis and Critical Control Points (HACCP) issues.

(c) Glue Application – Closure here is effected by applying adhesive to the surface of the inner flaps of a fibreboard case and pressing the outer flaps against them. This can be done either manually or with the aid of a machine. Effecting closure by gluing is the cheapest closure system known. However, if the glue quality is poor, setting time could be long and may result in flaps flipping back. To prevent this, cases are usually made to pass through a length of compression belt or stacked on top of one another after sealing to aid adhesive setting. If the bond strength is good, it can act as a tamper evidence device by revealing a fibre tear if opened by an unauthorized person.

There are no known disadvantages except that if not properly applied, glues can mess up the case and the contents.

Chapter 3 Packaging Materials and the Environment

Question 25 – What is a hermetically sealed pack? Describe the mechanism by which two such packs achieve the required shelf-life. Illustrate your answer by reference to different types of packs.

A hermetically sealed pack is one, which does not allow either air or water, including water vapour, to get inside. Examples of hermetically sealed packs follow.

Tinplate (metal) cans for such products as margarine, tomato puree, etc. The can is a three-piece combination, that is, the cylindrical body and the two circular ends. The tinplate itself – low carbon mild steel with tin coating on both sides – is impermeable to gases, moisture, etc. The body is formed by a folding side seaming operation either by soldering or welding. The two ends are mechanically joined to the cylinder by a double seaming operation in which two flanges or hooks of the body cylinder and end are mechanically interlocked. A gasket material or lining compound is applied to the end flange after stamping out, and this helps to produce a good hermetic seal in the formation of the double seam.

Tubes made of aluminium or other metallic material for toothpastes, creams, etc. are another example of a hermetically sealed pack. The tube making process involves forcing a slug of aluminium through a die to produce a work-hardened tube. After trimming, this is annealed to produce the familiar flexible soft tube. This could be further treated internally (lacquering) and/or externally (enamel coating) for various purposes. Since any container is as good as its closure, the tube cap, usually made of HDPE or PP, is well designed and resilient enough to produce a hermetic seal. After filling in the product through the open end, the latter is double folded and properly crimped mechanically. This packaging is good for shelf-stable products.

Other examples are products in glass containers, such as beer and carbonated soft drinks with cork crown closures carefully crimped on.

Question 26 – A biscuit is packed in a normally coated (MS grade) regenerated cellulose film and at the time of packing has initial moisture content of 2 per cent (limits +/- 0.5 per cent). Storage tests at 25°C. 75 per cent relative humidity (R.H.) gave the following weights:

	Initial weight	Weight after 50 days
Pack 1	210.5 g	212.8 g
Pack 2	211.0 g	213.0 g
Pack 3	214.0 g	216.6 g
Pack 4	212.3 g	214.7 g
Pack 5	209.7 g	211.8 g
Pack 6	213.1 g	218.7 g

The average weight (at equilibrium) of the packaging materials is 2.5 g. Assuming critical moisture content of 4.5 per cent of the product, estimate the shelf-life for:

(a) UK markets

(b) Tropical markets

Give reasons and comment on the experimental results. Indicate means of improving the shelf-life of the exported biscuits.

Ignore pack 6 because this is obviously a leaker.

Initial average net weight = 211.5 – 2.5 = 209 g (at 2% moisture).

Therefore, dry weight of product = 98% of 209 g = 204.82 g.

Moisture content at day 50 = 211.28 – dry weight = 211.28 – 204.82 = 6.46 g.

Therefore, moisture pickup from day 0 to day 50 is 6.46 − 4.18 = 2.28 g.

Since the critical moisture content is 4.5%, moisture gain up to the critical stage is (4.5 − 2.0) × 209 g, that is, 5.225 g.

2.28 g moisture was picked up in 50 days.

(2.28/50) g will have been picked up in 1 day.

Therefore, 5.225 g moisture will be picked up in (50/2.28)(5.225) days = **114.58 days**.

This will be the shelf-life for the UK markets at 25°C and 75% R.H.

For the tropical markets (38°C, 90% R.H.), the shelf-life will be (75/90) × 114.58 _= **95.48 days.**

The higher the R.H. gradient, the higher the rate of moisture pickup, and the relationship is nearly a direct one. This explains the reduced shelf-life obtained under tropical conditions. A better barrier film like the metallised regenerated cellulose may be required to obtain a reasonable shelf-life for the tropical markets.

Question 27 – 'Corrosion is not simply a matter of humidity'. Discuss this statement with particular reference to the packaging of an electric toaster.

This statement is true though many people wrongly believe that the humidity factor is all that is necessary for corrosion to take place. First, the relative humidity must be at least 50 per cent before any substantial corrosion can take place on corrodible materials such as electric toaster. Second, the presence of dust, dirt, or any particle of an electrolytic nature is necessary before corrosion can take place. Unfortunately, perfectly clean air does not exist anywhere since the air is always full of visible and invisible particles/gases, which in addition to the presence of moisture, do aid corrosion. This is why it is always advisable to rid corrodible materials, like metals, of dirt before packing them in equally clean packaging materials. Also, since it has been well established that high relative humidity in the presence of dirt accelerates corrosion, efforts should be made to ensure that the relative humidity within the package is not too high to prevent moisture buildup, hence corrosion. One method of keeping the relative humidity within a package containing corrodible materials/products low is the inclusion of silica gel, which is capable of absorbing moisture and therefore keeps the humidity low.

Generally, the necessary conditions for corrosion to take place are the simultaneous presence of

(i) Moisture

(ii) Air

(iii) Warmth

(iv) Dirt (dust, particles, etc.)

Question 28 – What is the difference between mould and bacteria? What are the conditions favourable for growth, and what steps can be taken to prevent or at least inhibit growth? Illustrate your answer with reference to the packaging of meat pies and disposable hypodermic syringes.

Bacteria are unicellular plants too small to be seen unless a microscope is used. They multiply by dividing in twos. One bacterial cell can make two in twenty minutes in the right conditions, and so the number increases rapidly.

Moulds on the other hand are spread in fungal pores, and they differ from bacteria in that when the pores grow, instead of simply dividing into two like the bacteria, they send out thin, root-like filaments (hyphae) which branch and grow.

The conditions necessary for bacterial growth are

(a) Availability of water

(b) Availability of food (organic substances)

(c) Right temperature (e.g. 20 to 30°C)

(d) Availability of oxygen (for aerobes) and non-availability of it (for anaerobes)

The conditions favourable for mould growth are almost the same as for bacteria except that moulds do not need liquid water to grow. Also moulds need oxygen for growth unlike the anaerobes that do not need oxygen.

Bacterial/mould growth could be prevented or inhibited as follows. Since moisture is the most fundamental need for both, the best way to prevent their growth is by water exclusion. This can be done by ensuring that products are packed in sealed moisture-resistant barrier to preclude condensation, provided the contents are dry enough (e.g. disposable hypodermic syringes). Damp products (e.g. meat pies), which are likely to release moisture, must be packed in porous or permeable materials.

Question 29 – Differentiate between water vapour, permanent gases and odours. Discuss the factors, which influence their permeation through flexible packaging media. Give one example of spoilage of a foodstuff caused by each diffusing medium.

Water vapour can be described as water in its gaseous form, and this can easily be condensed or liquefied back into water under high pressure at room temperature. Permanent gases on the other hand are those gases, which cannot be liquefied by the application of pressure at room temperature. Odours are volatile constituents of some products or compounds such as perfume or flavourings. While the compound may be solid, liquid, or powder, usually the volatile constituents are gaseous.

The conditions, which influence their permeation through flexible packaging media, are as follows: The first and most important factor is that no flexible packaging medium is a complete barrier because they all have at least some micro-pores that are wide enough to allow the individual molecules of water vapour, gases, or odours to pass through. Even in some laminates where aluminium foil is a component, the presence of pinholes in the foil where molecules could pass through cannot be ruled out.

Secondly, when there is a difference between the equilibrium relative humidity (E.R.H.) of a product and the relative humidity of its immediate environment, a sort of humidity or pressure gradient is set up, and this triggers the transfer of moisture from one location to the other through a flexible permeable medium or membrane. This transfer will continue until equilibrium is reached or established.

Finally both the temperature and pressure influence the permeation of gases or vapours through flexible media except the permanent gases, which are hardly affected by pressure.

Examples of food spoilage:

(i) Diffusion of oxygen, O_2, into fat-based products such as margarine results in oxidation processes which cause rancidity and eventual emulsion breakdown.

(ii) The ingress of water vapour into sachets of instant powder drinks and savoury products results in product caking which renders the product useless and unacceptable to the consumer.

(iii) The loss of SO_2, a preservative in some squash drinks results in product deterioration and eventual off-flavour taste.

Question 30 – (i) The forces which act on a body brought rapidly to rest are related to two main factors. What are they?

(ii) Explain how a deceleration of 15 g acting momentarily on an object can cause breakage.

The forces acting on a body brought rapidly to rest are related to the following two main factors:

(a) The mass of the body on which the forces act
(b) The magnitude of the decelerating forces

The more rapidly an article is brought to rest, the greater the shock it experiences. Shock is the force or stress resulting from a sudden change in velocity. Every article dropped from a given height (drop height) is associated with some shock value called the G-factor or the impact-load factor above which the object will break and below which the object is considered safe. This deceleration can be determined experimentally. The cause of damage is due to sustained shock, which leads to displacing some part of an object beyond its elastic limit. After such displacement, the article will not return to its original position. In respect of the example above, the 15 g deceleration acting even momentarily can cause breakage if the 15 g is greater than the G-factor of the object. If at all 15 g is less than the nominal G-factor of the object, breakage can still occur if the duration of the shock is too long for that object.

Question 31 – State and briefly explain the three principal ways of preventing shock damage.

The three principal ways of preventing shock damage are

(i) Distribution of shock

(ii) Localisation of shock

(iii) Cushion or shock absorption

(i) Distribution of shock – When forces of equal magnitude are applied to two surfaces of unequal areas, the effect of the forces is felt more on the surface with the smaller area than the other surface. In short, what is really important is the force per unit area, that is, pressure. Hence, in order to reduce shock to a safe level, the area over which the shock forces act is increased, thus distributing the force and minimizing its effect.

(ii) Localisation of shock – This could be regarded as the opposite of the above method. Localisation involves restricting the shocks to certain points to prevent other points from receiving the shocks and so prevent damage. Usually the forces at impact are directed to the stronger points of the product/package, which are capable of accommodating the shocks, thereby preventing damage to the weaker areas of the product/package. This method of prevention is particularly suitable for large equipment, such as aircraft engines.

(iii) Cushion or shock absorption – This is a method whereby the energy of the body is absorbed by another body (cushion) and the former is brought to rest gently. Cushioning is the employment of material to mitigate shock. Three different classes of material are usually employed for providing cushioning systems, and these are

(a) Resilient cushions

(b) Non-resilient cushions

(c) Space fillers

Question 32 – Explain how to measure cushion characteristics.

One method of measuring the cushion characteristics is the use of a load-compression curve. This is obtained by taking a piece of cushioning material, loading it to various levels, and plotting the amount by which the cushion is compressed at each particular load level. From the readings taken, a load-compression curve, as shown below, is produced.

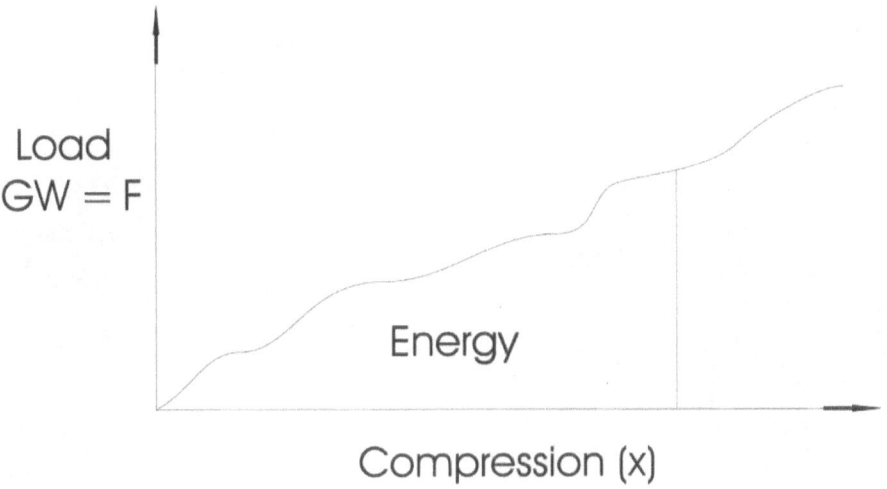

Figure 3.1 Load-Compression Curve

The energy corresponding to any particular compression is given by the area under the F-x curve. By numerically integrating the F-x curve, another curve, the load-energy curve below can be obtained.

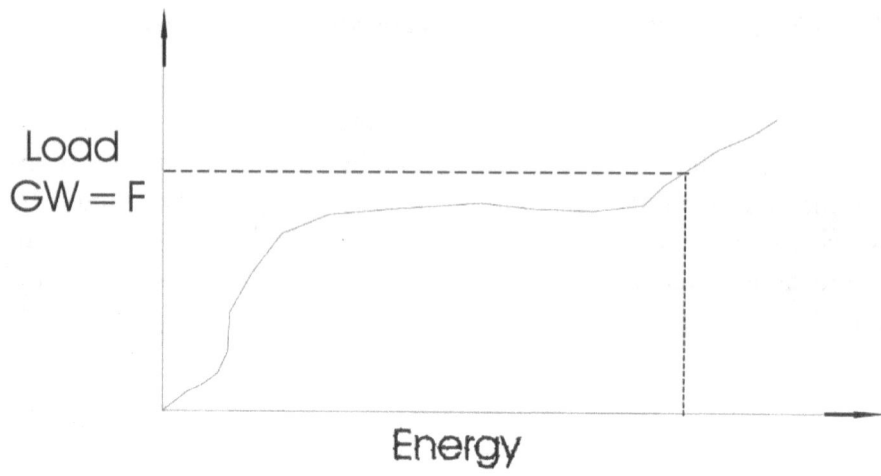

Figure 3.2 Load-Energy Curve

We can now determine the cushion factor, C, from the relationship C = Load/Energy, and the relationship between C and load can be drawn as follows.

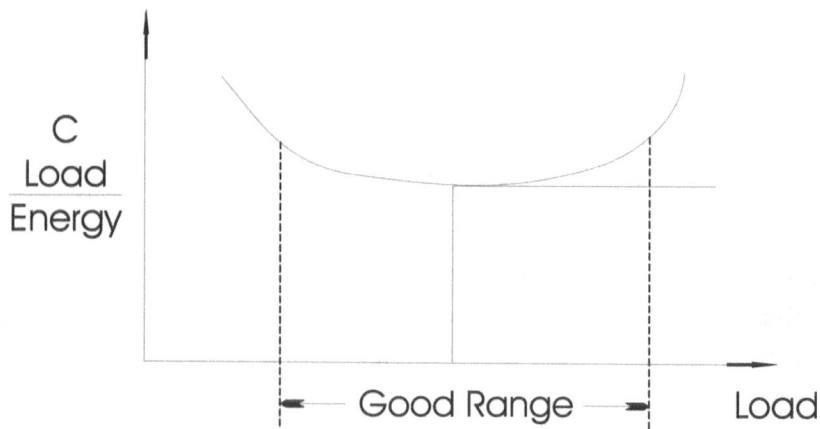

Figure 3.3 Load Energy – Load Curve

From the above curve the load characteristics at which the cushion works most efficiently can be determined, and this is the region where C is at the minimum. Fortunately the C-load curve is rather flat at the bottom, that is C is not very sensitive to load near the minimum C value, which is an advantage in design work.

Question 33 – What are the six steps in rational cushion design? If package A is dropped onto package B from a height of h, to what physical forces are the two containers and their contents subjected, and how do these forces arise?

The six steps in rational cushion design are as follows:

(1) Assess the distribution hazards and decide on design drop height and number of drops

(2) Assess product fragility

(3) Choose possible materials that will provide the required protection and calculate amount of each one needed to provide it

(4) Calculate total distribution costs

(5) Make prototype packages using best material

(6) Assess prototype packages

Package A is acted on by the force of gravity due to its own weight/ mass. If the mass of package A is W_A kg, then a force of $W_A \times g$ kg acts on it in the vertical direction, where g is the acceleration due to gravity.

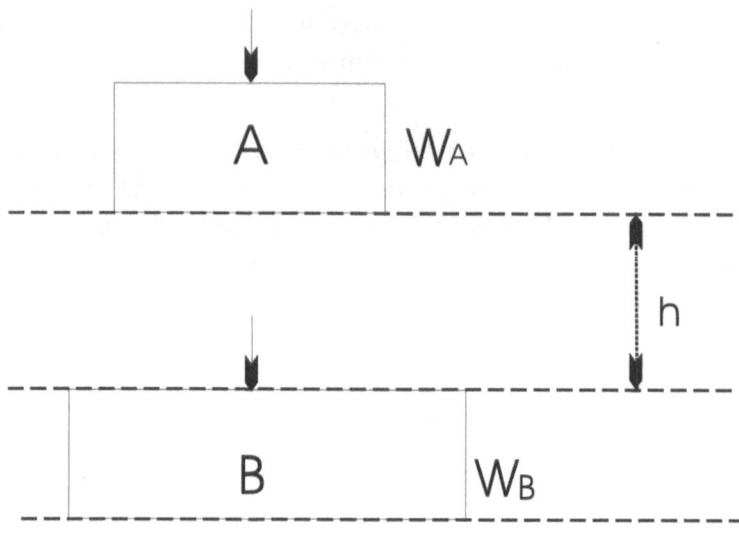

Figure 3.4

The potential energy (P.E.) $W_A \times h$ possessed by A at the height h will have been fully converted to kinetic energy (K.E.) just before impact with package B. After impact this energy will be absorbed by package B.

Package B and its contents are acted on by the combined forces due to its own mass and that of package, A which falls on it. These forces all have gravitational origin.

Chapter 4 Paper and Paperboard Packaging

Question 34 – What defects in wood for packaging must be avoided and why? Why are top quality soft woods little used for making timber packages? For what purposes is top quality timber essential?

Woods for packaging must not have the following defects:

(i) Excessive moisture content. Woods must be seasoned by drying to around 15 to 20 per cent moisture content to reduce the magnitude of dimensional changes due to shrinkage and swelling. Moisture control also protects the wood from attacks by micro-organisms, reduces the weight, and better prepares the wood for most finishing and preservation methods. Finally, it increases its strength.

(ii) Woods must not contain defects which materially reduce their strength (e.g. bad cross grain, knots greater than one-third the width of the pieces concerned, or knots that run into the edges). Woods should be free from wane. The presence of all these defects reduces the strength of woods in whatever use they are put.

Top quality soft woods are little used for making timber packages because these woods are no longer plentiful and are now very expensive. Demand almost outweighs supply. Therefore cheaper grades of timber are now used for box making. Top quality woods or timbers are used only when their strength characteristics are needed to carry their contents safely to their destinations or when the value of the contents warrants the extra protection.

Some of the uses for top quality woods are furniture and building works (doors, windows, etc.).

Question 35 – What are five primary sources of fibrous material for paper and board making, and what is the most important one? What are two non-fibrous materials, and what are their uses? What does wood contain in addition to fibrous material and in what amounts?

Five primary sources of fibrous material for paper and board making are

    (i) Wood pulp

    (ii) Cotton

    (iii) Flax and esparto

    (iv) Bamboo

    (v) Straw

Wood pulp is the most important one.

Two non-fibrous materials and their uses are

    (i) Clay – This increases the opacity and brightness of paper and boards and also improves their printing properties.

    (ii) Rosin – This is a sizing agent and makes the end product (paper and board) water-repellent.

In addition to fibrous material, wood contains the following materials and in the proportions indicated.

Material	Proportion (%)
Lignin	30
Carbohydrates	16
Proteins, resins, and fats	4

Question 36 – What are the three most important differences between the pulps produced from spruce wood using the following processing methods:

(a) Mechanically by grinding

(b) Chemically by the kraft process?

The three most important differences between the pulps produced from spruce wood mechanically by grinding and chemically by the kraft process are as follows:

(i) In the kraft process, the fibres are long, and the pulp yields stronger paper than is produced by mechanical grinding. The kraft process produces purer cellulose.

(ii) The cellulose fibres produced by the kraft process are of higher quality than the ones produced by the mechanical grinding process. In mechanical grinding, more fibre bundles emerge, and the damage to fibres are greater than in the kraft process.

(iii) In the kraft process, all materials except the cellulose fibres are dissolved in solution by the chemical action, and the dissolved materials can be washed away, whereas in mechanical pulping, the water soluble impurities are the only ones which are removed and most of the lignin still remains.

Question 37 – (a) Long fibres are considered useful in imparting strength to paper. What other effects can long-fibred pulps have on the properties of the sheet?

(b) State three factors related to the fibres of the pulp and which determine the strength of a sheet of paper made from it. How does the paper maker attempt to control them?

(a) The effects long-fibred pulps can have on the properties of the sheet are

    (i) Blotchy look

    (ii) Greater stretch

    (iii) Better wet strength

(b) Three factors related to the fibres of the pulp and which determine the strength of a sheet of paper made from it are

    (i) Length of the fibres

    (ii) Fibrillation

    (iii) Fibre clumps

The paper maker attempts to control these factors by a careful handling of the beating process, or fibrillation of the fibre bundles. This is because the beating process is a double-edged sword as far as its effects on paper characteristics are concerned. The extent of the beating process is a function of which of the paper characteristics is of interest to the paper maker and/or the end user. For example, a small amount of beating results in papers with high tear strength, high absorbency, lower burst strength, and lower tensile strength. On the other hand, increased beating reduces tear strength and absorbency but increases burst and tensile strengths. In short the paper maker has to perform a balancing act in order to achieve the desired end result.

Question 38 – Why can thicker paperboard not be made on single-wire paper machines? How do machines using two wires or cylinders overcome the difficulty? How are the surface properties of the finished sheet affected by using two wires instead of one?

Thicker paperboards cannot be made on single-wire paper machines because after the thickness of the mat (or paper in the making) has reached a certain limit, draining difficulties arise because most openings, if not all, are now blocked. This phenomenon can be compared to a filtration operation, which cannot continue much longer once the thickness of the cake has reached a certain level.

The drainage problem is overcome by using a two-wire (cylinder) paper machine. The idea is to form the paper web rapidly, usually between two wires, which could be horizontal or vertical so that drainage can take place from both sides.

The use of two wires instead of one makes the two surfaces of the finished sheet similar.

Question 39 – State two main reasons for giving the surface of paper or board a finishing treatment. Describe one treatment in broad outline and give details on how the quality of the treatment is measured in practice.

Two reasons for giving the surface of paper or board a finishing treatment are
- (i) To improve the resistance of the finished product to water, water vapour, gases, greases, and oils.

- (ii) To increase opacity and brightness and to improve the printing properties.

Sizing/coating by wet waxing – This treatment is suitable when a film of wax on top of the sheet is desired. After the wax film has been applied to the surface of the sheet, it passes into a bath of cold or refrigerated water, which immediately sets the wax before it has enough time to penetrate into the sheet.

Since the purpose of this treatment is to improve the resistance of the material to liquids and gases, the best way to measure the quality of the treatment is to determine the rate of water vapour permeability of the treated material using an untreated piece as control. This test can be done using the B.S.I Dish Technique.

Question 40 – Define setting time and tack. Discuss the influence, which these properties have on the performance of adhesives on a modern packaging line. Give examples to illustrate your answer.

Setting time is the time required for a fibre tear to be achieved between two substrates from the time they are bonded together by an adhesive.

Tack is the initial resistance offered by an adhesive at the wet stage.

Setting time and tack are very important to the performance of adhesives on a modern packaging line in that these lines are very fast with some producing as much as several hundred packs per minute. The time from the point of adhesive application and sealing of a packet to the time of putting the packets in a shipping container is very short. But at the same time, setting is supposed to have taken place by the time these packets are put in the shipping containers so that no product leakage occurs. It is therefore essential that the setting time should not be unduly long. A good example is the packing of detergent powder in wax or non-wax laminated chipboard cartons. The success of this operation, as far as ensuring leak-proof packets is concerned, depends on the setting time of the adhesive. However there are other factors, which affect adhesive setting time adversely (e.g. inadequate conveyor belt pressure, poor adhesive application).

Tack becomes very important in bottle labeling operations. Most times, immediately after bottles are labeled, they are packed manually into outer cases by operators before setting time. In the process, the operators' hands and the case walls (or its internal dividers) make contact with the labels on the bottles. If the tack is not good enough, squeezing, skewing, or wrinkling of labels may occur, which may result in poor product presentation.

Question 41 – What are machine direction (MD) and cross direction (CD)? Why are they crucial in the design of cartons and labels to be used on automatic packing/labeling machines?

During paper/board making, a fibre and water slurry is deposited onto a moving belt. The fibres tend to align in the direction of travel, known as the machine direction (MD) or grain direction. The direction across the paper making machine or across the fibre alignment is known as the cross direction (CD).

The fibre alignment gives paper and board different properties in the MD and the CD directions. For example, paper and board tear and fold more easily along the MD than along the CD. On the other hand, stiffness values of paper and board are higher in the machine direction than in the cross direction. Depending on the type of board making machine, the stiffness value ratio MD:CD ranges from 2 to 4 for Fourdrinier- and Cylinder-made boards, respectively.

The implications of these different properties on automatic packing machines are as follows. In producing cartons for use on an automatic packing machine, blank layout must take the grain direction of the board into account. To avoid bulging when the carton is filled with certain products (e.g. granular powder) and to ensure hitch-free carton erection, the grain must run around the carton perimeter.

For paper meant for labeling purposes, the paper must not curl. While the grain direction of the label is usually specified by the labeling machine manufacturer, the direction is usually (but not always) parallel to the copy, that is, it is parallel to the axis of curvature of the container on which it is applied.

Question 42 – What is tinplate? What is tin-free plate? Describe briefly, the manufacture of tinplate from steel ingots. What are the main advantages of tinplate as a packaging material? What differences in can manufacture are dictated by the replacement of tinplate by tin-free steel?

Tinplate is low carbon mild steel coated on both sides with tin. Tin-free steel is mild steel plate, electrolytically coated on both sides with chromium/chromium oxide. First, the base steel or strip is made by the rolling of hot steel ingots down to a strip with a thickness of 1.8 mm. This strip is then pickled continuously in a bath of hot dilute sulfuric acid to a finished gauge of 0.15 mm to 0.50 mm. The sheet is finally annealed and temper rolled to impart the required hardness and surface finish.

Next is the tin coating which can be applied either by running the steel plate through a bath of molten tin (hot dipping process) or by a continuous electroplating process.

The main advantages of tinplate as a packaging material are as follows:

(i) Tinplate is corrosion resistant.

(ii) Tinplate offers a hermetically sealed pack and, as a result, protects the product it contains over a long period of time without appreciable product deterioration.

(iii) Tinplate is strong and rigid, hence cans made from it can withstand rigorous handling during warehousing and distribution.

(iv) Tinplate can be easily decorated by modern printing methods (e.g. lithographic process), hence the need for labeling may not arise.

When tinplate is replaced by tin-free steel, can making by side soldering is no longer possible. In place of soldering, side seaming is effected by one of the following methods:

(i) Welding

(ii) Cementing using plastics (e.g. nylon)

(iii) Side lock

Fortunately soldering is no longer popular, particularly with the food processors because of the health dangers posed by the solder composition, which is a mixture of lead and zinc.

Question 43 – Describe how an open top metal can is made.

The open top metal can is made from tinplate, that is, low carbon mild steel plate coated on both sides with a thin layer of tin. To manufacture this style of can, a rectangular blank of tinplate is mechanically wrapped around a mandrel to form a tube or cylinder. The body blank is sized to be capable of making one can body. Before forming the tube, the body blank is stamped out at the four corners, where specially shaped notches are created so that instead of the interlocked seam being formed over the whole length of the can body, the extreme ends are simply overlapped. During body forming where the two ends of the blank meet, an interlocked seam is normally formed, and this is soldered to make it leak-proof. The side seam can also be secured by the welding process. Next the bodies (cylinders) are transferred to a flanging machine where the rims (the two ends) are flanged outwards to enable the ends to be seamed on.

The lids or ends are produced or stamped from similar sheets of tinplate. To avoid leakage, a resilient (lining) compound is applied to the whole peripheral area of the ends before seaming takes place. The end is then mechanically joined to the cylinder by a double seaming operation in which two flanges or hooks of the body cylinder and the end are mechanically interlocked. The can maker applies one end only, the other being applied by the food processor after product filling.

The blank sheets constituting the body cylinder are normally printed and dried before the body forming operations and the printing of plates is usually done using the offset lithographic method.

Question 44 – A typical food can might be described as being a '307×406'. What does this mean?

The conventional format for can containers adopts a three-digit number for the can dimensions. For a cylindrical can, two dimensions are normally specified, that is, the diameter and the height, in that order. For example, if a can is described as being a '307×406', the interpretation is as follows: The number '307' represents the diameter (outside seam to outside seam). The first digit, '3', denotes the whole number in inches, while the last two digits, '07', represent sixteenths of an inch. The second set of three digits, '406', is the height of the can, and the first digit, '4', denotes the whole number in inches, while the last two digits, '06', denote sixteenths of an inch. Therefore, 307 × 406 denotes a can of diameter $3\,^{7}/_{16}$ and a height of $4\,^{6}/_{16}$ inches. These dimensions are now largely being replaced by their metric equivalents in millimeters, that is, 307 × 406 will be replaced with 87.3 mm × 111.1 mm.

Question 45 – Describe briefly, two methods of making rigid containers from aluminium. Why can't aluminium cans be made by the method used for making tinplate cans?

Two methods of making rigid containers from aluminium are

(i) Impact extrusion

(ii) Drawing or stamping from sheet

(i) Impact extrusion – A circular disc of metal called slug is heat treated to bring it to the desired metallurgical state for fabricating. The slug is then lubricated and rammed through a die to produce a work-hardened container. The container wall is then cut to the correct length and either flanged for subsequent seaming, given a threaded profile to take a screw cap, or given an external bead profile to take a plastic snap-on cap. The containers could either be lacquered or anodized after manufacture in order to protect them from corrosion.

(ii) Drawing or stamping from sheet – The sheet which could be pre-lacquered is fed into a press (or series of presses) where a metal blank is cut out, formed to the shape of the can, and finally trimmed. This process is used for such cans as the rectangular fish can, circular paste cans, or for the containers used to hold fifty cigarettes. Deeper drawn cans are manufactured in stages (e.g. drawing and ironing process).

Aluminium cans cannot be made by the methods used for making tinplate cans because they cannot be soldered by any commercially available technique.

Question 46 – What are the properties of aluminium foil that make it attractive for use in flexible packaging? What are the limitations of aluminium foil, and how may they be overcome?

The properties of Aluminium foil which make it attractive in flexible packaging are

(i) Decorative properties – Aluminium foil is easily printed on or embossed.

(ii) Barrier properties – At thicknesses of about 25 microns and above, aluminium foil is almost totally impermeable to gases, water vapour, odours, oils, fats, and greases. This is also true at lower thicknesses if foil is laminated to other flexible materials like paper and plastic films.

(iii) Dead fold characteristics – Aluminium foil folds easily and retains creases once they are formed. This is an important property when considering using aluminium foil on packaging machinery.

(iv) Sterility property – The high temperature employed while annealing the foil destroys all micro-organisms, and provided there are no fingerprints present, the foil is completely sterile.

(v) Radiation property – If the bright side of the foil is on the outside of the package, aluminium foil is a good reflector of heat and as such keeps its content at constant temperature.

(vi) Another good property of aluminium foil is the ease with which it forms laminates with other flexible packaging materials, such as paper or plastic films.

The limitations of aluminium foil are corrosion and the mechanical sensitivity of unsupported aluminium foil. Both defects can be corrected by laminating the foil to paper or some other flexible substrates, like plastic films.

Question 47 – What is the electromotive series, and what is its importance to the study of corrosion?

When metals are arranged in the order of the ease with which they go into solution or form ions by electron release, then the resulting series is known as the electromotive series (EMS).
The importance of E.M.S. to the study of corrosion can be explained as follows. The ease with which metals go into solution is a measure of metal corrosion taking place. Therefore metals at the top of the series (e.g. Na) go into solution comparatively easily, while those at the bottom of the series (gold) do not go into solution and so do not corrode. Metals that are far apart in the series produce a greater voltage when

they form the electrodes in a galvanic cell, so the one that is the anode (the one nearest to the top of the series) will corrode rapidly. Therefore, by a careful study of the EMS, one is able to make the right selection of metals to be used for electrodes at a given situation to achieve a desired goal. The same consideration goes for the marriage of two or more metals in fabrication processes.

Question 48 – What is a galvanic cell? Explain how a galvanic cell is formed when a liquid is filled into a tinplate container. What is the basic mechanism of corrosion?

When two dissimilar metals are immersed in a conducting solution (or electrolyte), then we have a galvanic cell. An electrical potential is set up between the two metals and current will flow if they are connected by a wire.

When a liquid is filled into a tinplate container, a galvanic cell is formed as follows: The tin component of the tinplate constitutes an electrode (usually the anode). So basically the tinplate constitutes a pair of dissimilar metals (electrodes) in the liquid (presumed conducting). Thus a galvanic cell is formed.

The basic mechanism of corrosion is electrochemical involving a transfer or displacement of electrons.

Question 49 – What are two factors, which must be present for corrosion of iron to occur? What other factors affect the rate of corrosion?

The two factors which must be present simultaneously for corrosion of iron to occur are oxygen and moisture.

The other factors that affect the rate of corrosion are

(i) Polarization is the build up of hydrogen on the cathode. This reduces corrosion.

(ii) Aeration of the environment (e.g. high oxygen concentration). This makes the tin cathodic, thus increasing the corrosion rate.

(iii) The materials from which the electrodes are made.

(iv) The pH of the environment.

(v) The temperature of the environment. Since corrosion itself is electrochemical in nature, like most chemical reactions, the corrosion rate increases with increase in temperature.

(vi) An unclean environment. This also increases the rate of corrosion.

Question 50 – How can corrosion be reduced or eliminated? Give examples of how they are used commercially in packaging.

Corrosion can be reduced or eliminated in a number of ways. First, no corrosion will take place if all access to liquid water is prevented and relative humidity of the atmosphere inside the package is kept below critical humidity for the metal concerned. To achieve this, barrier materials now exist that can reduce the amount of moisture that gets into the metal surface to a safe level. In addition to this, the use of desiccants (moisture-absorbing agents) to keep the moisture at bay is common.

Second is the surface protection of corrodible materials. Protecting the metal surface is probably the oldest of the procedures for corrosion control. A painted surface, for example, isolates the underlying metal from the corroding electrolyte. The chief limitation of this method is the service behaviour of the protective coating.

Inhibitors have been developed as a way of capitalizing on the tendency for large polyatomic anions to adsorb onto a metal surface. The basis on which they work is by interfering with the functions of the electrolyte or electrodes, which has the effect of slowing down or stopping the corrosion process.

Another way of reducing or eliminating corrosion is by avoidance of galvanic couples, and the simplest way of doing this is by limiting designs to only one metal. Apart from the fact that this is not always feasible, a galvanic couple may even develop in a single material because of microstructures, stress concentrations, or electrolyte heterogeneities.

Some examples of corrosion reduction or elimination processes in commercial use in packaging are

(i) Aluminium foil laminated to paper or film prevents corrosion of the former.

(ii) Tin coating in tinplates serves the same purpose.

(iii) Chromium/Chromium oxide coating in TFS.

Question 51 – How can a coating of tin reduce the corrosion of steel, even if the coating is incomplete and contains pinholes?

Tin-coated steel is what has been defined as tinplate. In short, tinplate is the marriage of two dissimilar metals, that is, tin and steel. The combination therefore constitutes a galvanic cell with the product packed in the tinplate container acting as the electrolyte. The tin and the steel constitute the electrodes. Ironically, in most cases, the tin acts as the anode and therefore dissolves in preference to the steel component, which now acts as the cathode. Since it is the dissolution of metal that is referred to as corrosion, steel is preferentially protected even if there are pinholes exposing steel to possible attack.

Chapter 6 Glass Packaging

Question 52 – What are the principal constituents of glass? What is cullet, and why is it important to glass manufacture? What are the two methods of moulding glass containers? What type of container is most typically made by each moulding method?

The principal constituents of glass are

Sand (silica) – SiO_2,	73%
Soda Ash – Na_2O,	14%
Lime – CaO,	11%
Alumina – Al_2O_3,	1%

Other minor ingredients make up the balance of 1 per cent.

Cullet is scrap or broken glass recovered from the plant operations or the trade, and it can constitute a substantial proportion of the raw materials feed into the furnace. Its presence enhances the melting rate and significantly reduces energy requirements.

The two methods of moulding glass containers are

(1) Blow and blow process
(2) Press and blow process

The blow and blow process is typically used for the moulding of narrow-necked bottles, while the press and blow process is used for the production of wide-mouthed jars.

Question 53 – What are the main advantages of glass as a packaging material? What steps have been taken to overcome the disadvantages?

The main advantages of glass as a packaging material are

(a) Chemical inertness

(b) Barrier properties

(c) Clarity or transparency

(d) Rigidity

(e) Resistance to internal pressure

(f) Heat resistance

(g) Low cost

The main disadvantages of glass as a packaging material are

(a) Fragility

(b) Weight

Fragility in a glass container could be described as its tendency to break, crack, or chip. Even at the design stage, efforts are made to prevent or reduce this defect by paying attention to shape and glass distribution. During filling, when most cracking incidents occur, precautions, such as cushioning of conveyors and guide rails, are taken to avoid cracking. Also distribution partitions are used to keep bottles apart in the corrugated outer cases so that cracking does not occur as a result of bare. body contact/impact.

The problem of weight is not unconnected with fragility because early glass containers were made very heavy in order to have enough strength to withstand handling on the packing lines and in distribution. However, nowadays successful efforts to reduce the glass weight while still ensuring good strength include

(1) Better design

(2) Better glass weight distribution

(3) Chemical treatment at the annealing stage so that one can use less glass and still be sure of a strong container. One such treatment is titanising (i.e. the spraying of titanium oxide onto the glass surface while still hot). This treatment has been found to reduce the development of micro-cracks and increase the lubricity of the glass surface.

Question 54 – What is thermal shock? How is it important to users of glass containers? In what type of filling operation is resistance to thermal shock important?

Thermal shock is the effect of sudden temperature changes on glass containers. The effect is minor without any undesirable results if both surfaces of the glass suffer the same sudden temperature change. The problem of thermal shock becomes uncontrollable when only one surface is subjected to a sudden temperature change. This is because if one surface of a glass bottle is rapidly cooled, contraction takes place faster in that region of the bottle while the opposite surface hardly shows any contraction. This uneven contraction sets up stresses on the bottle with the possibility of a terrible crack.

This phenomenon is very important to users of glass containers because most of the packaging/sealing/processing operations in which glass bottles are used involve the use of fairly high temperature treatment, which may have adverse cracking effects on the bottles if care is not exercised. Some of the precautions that could be taken to avoid thermal shock are

(i) Controlling the thickness of the bottle

(ii) Avoiding abrupt (sharp) joints at radiused areas of the bottle

(iii) Keeping sudden temperature drops to a reasonable range

Resistance to thermal shock is important in any filling operations involving

(1) Filling of glass containers with a hot product

(2) Cooking or sterilization of product in the container

(3) Sterilization of empty containers by steam or dry heat

Question 55 – What are the main features of a good bottle closure?

The main features of a good bottle closure are as follows:

(a) It should prevent loss of the contents or any constituents of the contents; for example, it should prevent loss of perfume from a hand cream.

(b) It should also prevent penetration of any substance from outside of the container.

(c) The closure material should not react in any way with the contents of the container, that is, there must be product/closure compatibility.

(d) It should be easy for the consumer to reach the content.

(e) It may have to make a good re-seal (or re-closure) as with the closure on a jar of instant coffee, this being a hygroscopic product.

(f) It should harmonize with the container. It should add to the sales appeal of the product.

(g) It may have to be tamper-evident.

(h) It may have to be child resistant.

(i) It may have to act as a dispenser.

Question 56 – What are the main resilient materials for sealing glass containers? What are their advantages and disadvantages?

The main resilient materials used for sealing are: corks, rubber, wood pulp, plastics, and flowed-in compounds.
Their advantages and disadvantages are as follows:

Cork – Advantages are (i) It is very resilient and does not burst, (ii) it is inert and can therefore be used in contact with a wide range of products.
The disadvantages are that it can impart some taste to the product over a long period of time and the material is brittle and can easily crumble.

Rubber – The advantage is that it is very resilient. The main disadvantage is that it imparts a strong flavor to foods or beverages, and it is therefore not generally suitable.

Wood Pulp – No obvious advantage is known apart from the fact that it is cheap. The disadvantages are that (i) it is not very resilient, (ii) it can impart a taste to sensitive foodstuffs unless wax coated when used in contact with liquids.

Plastics – The advantage is that there is little or no odor transfer when in direct contact with the contents. The disadvantage is that it has little resilience except when plasticised.

Flowed-in compounds – These may be based on PVC or rubber compounds. Usually they are melted and then injected into the peripheral channel of the upturned closure. These have the advantage of needing no separate sealing material that could be lost during use. No major disadvantage is known.

Question 57 – What methods are available for fitting screw closures to bottles?

The three methods available for fitting screw closures to bottles are as follows:

(i) By manual method – Screw closure can be fitted and tightened by hand. The disadvantage here is that no consistent result can be obtained, and this can result in improper seal.

(ii) By semi-automatic method – Here the cap is put loosely on the bottle. The bottle is then held by hand under a revolving head, which descends, grips the cap, and tightens it.

(iii) By fully automatic machines – Here the caps are fed to the bottles from a hopper and the bottles conveyed to a single or multi-head tightening head. Consistent cap tightening is obtained by using a friction clutch in the capping head. Here the advantages are enhanced productivity and consistency of tightening torque.

Question 58 – The term 'Annealing' is often encountered in the packaging literature. State three different packaging processing areas where the term is used and explain its importance in each case.

The term annealing is generally used to denote either a softening or a toughening process. In the glass making technology, annealing is used to describe a heat treatment for removing residual stresses and thereby reduce the probability of the development of thermal cracks in a brittle glass. Hence, in this situation, annealing denotes a toughening process.

Annealing is also a crucial stage in the steel plate making. Here annealing takes place at a temperature of about 650ºC and this results in a marked increase in plate ductility and a decrease in strength. Therefore annealing in steel plate making is largely a softening process.

Finally, annealing is vital in the making of collapsible aluminium tubes. When the tubes are formed by the impact extrusion process, they are hard and stiff. The annealing process that follows softens the tubes and makes them flexible and collapsible for the tubes to function properly.

Chapter 7 Packaging Machinery and Related Issues

Question 59 (a) Which of the four basic types of liquid-filling machines is best for fast filling? Which is the cleanest and most economical way of handling most products?

(b) Which is the most satisfactory way of meeting the requirements of the Weights and Measures Act when dry goods in granular form are to be filled?

(a) Vacuum filling is the best for fast filling of liquids. Also vacuum filling is the cleanest and most economical way of handling most products.

(b) Filling by weight, that is, gravimetric filling, is the most satisfactory way of meeting the requirements of the Weights and Measures Act when dry goods in granular form are to be filled.

Question 60 – State the three main operations to be performed in any cartoning system. Illustrate each of these with an example from a product with which you are familiar. What other operations may have to be performed on a cartoning production line?

The three main operations to be performed in any cartoning system are

(a) Forming, or erecting, the carton or container

(b) Filling, or loading, the carton or container

(c) Sealing, or closing, the carton or container

The three operations above are common with the packing of a detergent powder such as Omo, Drive, etc. The cartons, usually delivered flat, are fed into the machine magazine in an upright flat position. Forming, or erection, is done automatically by suction on the line. Polyvinyl Acetate (PVA) glue or any other suitable adhesive is applied to the bottom flaps and sealed while the erected carton is being conveyed to an overhead hopper where filling of the powder into the container is carried out through a discharge chute. This is finally followed by adhesive application to the top flaps and eventual closing or sealing. In actuality, this closure/sealing operation takes place at two stages: first at the bottom flaps prior to filling and second at the top flaps after filling.

Other operations, which may be performed on a cartoning production line, are insertion of leaflets or any promotional items into the cartons. These could be done manually or automatically. Also date code markings could be added as well as over-wrapping of the carton with films, such as BOPP, as a tamper evidence device. In addition to tamper evidence, this latter operation also improves product presentation in the market place.

Question 61 – Outline the ways in which glass bottles may be delivered from the bottle manufacturer to the user's filling line, giving one advantage and one disadvantage of each method from the point of view of

    (a) the bottle manufacturer
    (b) the user

Glass bottles may be delivered in several ways from the bottle manufacturer to the user filling line.

   (i) Delivery of bottles in corrugated outer cases – Usually this is done by partitioning the outer case into cells or compartments equal to the number of bottles in a case. The partitions prevent the bottles from making contact with one another and therefore reduce the chances of breakage. Also, in a situation where the user does not do any pre-washing of the bottles before use, the bottles may have to be capped at the manufacturer's end before they are packed into the outer cases. Most of the time, the outer cases are later used at the filler's end to pack the filled bottles for warehousing and eventual distribution.

The above method is a rather costly and poor method of bottle delivery. It subjects shipping containers to multiple handling, which results in the weakening of the containers, particularly by the time the product is in distribution. This may call for the use of an expensive higher grade of board. The capping operation at the manufacturer's end slows down rate of production. Bottles in outer cases take a lot of storage space. Finally the capping operation at the manufacturer's end will have to be undone at the filler's end before feeding the bottles into the packing line. This normally calls for additional labour and handling, which could result in bottle contamination.

(ii) Palletized delivery system – This is the most popular and efficient method of bottle delivery, and many variants of it exist. The type chosen by any supplier/user depends to a large extent on the level of cleanliness required and the facilities at the disposal of both the supplier and the user. For example, delivering glass bottles in palletized format to a user who has no de-palletizing facilities makes no sense from the operational point of view.

Generally bottle palletization consists essentially of placing a fibreboard tray or layer on a pallet on which the bottles are arranged. The length, width, and sleeve (if any) of such trays depend on the bottle size and the number of bottles that can be accommodated. The use of a top tray/layer is optional and depends on the wishes of the individuals. There could be four to eight layers of bottles forming a full pallet depending on the bottle height and weight. Where adequate handling facilities are available, the pallet height rather than the weight becomes the limiting factor. Next is the shrink-wrapping or stretch-wrapping of the pallet assembly. This is followed by strapping to firm up the pallet pack. The number of straps used depends on the size of the pallet and the level of stability expected.

The advantage of bottle delivery in outer cases to both the manufacturer and the user is its simplicity and low or no capital outlay on the part of either party. A disadvantage to the manufacturer is the slow down in operations resulting in low efficiency. The disadvantage to the user is the multiple handling of the cases, which could result in the weakening of the case and its crushing unless a higher grade of board with its attendant higher cost is used.

The advantage of bottle palletization to both the manufacturer and the user is that palletization provides efficient utilization of storage space. A disadvantage of it for both is the inevitable investment in handling equipment, though this is not much of a financial burden for a high volume bottling operation.

Question 62 – Four hundred thousand 4-oz glass bottles of a proprietary liquid medicine are to be distributed in machine-packed cartons per annum. List the points needing consideration in the choice of carton style and method of production. In the light of these points, make your decision as to what conclusions you would reach for this particular operation.

In making the choice of a carton for the bottles, the following factors will need to be considered.

The shape and size of the bottle will be an important factor since this will determine the basic dimensions of the carton. Also the weight of the bottle and its content will suggest what grade of board will be suitable for the carton since the carton is expected, among other functions, to offer some degree of protection to the glass bottle.

For the carton style, we need to consider not just the carton in relation to the packing machine, but also what happens when the product gets to the ultimate consumer. Easy opening of the carton and easy re-closing of it each time the product is used until the bottle content is exhausted requires that the carton be the tuck-end type. Also the need to prevent the base flaps from giving way during use, thereby resulting in bottles falling off the base, calls for crash-lock carton base design.

The obvious method of carton production is the offset lithographic printing method. If the volume and quality of the design justify it, the gravure method of printing could be adopted. Either way, printing will be followed by cutting and creasing of the printed sheets on a die-cutting and creasing machine, called 'forme'. After printed blanks have been cut and creased, the surrounding waste, as well as any internal waste between or within the blanks, must be removed. This operation, known as 'stripping', is carried out partly or wholly on the press itself, by hand, or by mechanical hammers as a separate operation away from the press. Finally the blanks are fed into the folding/gluing machine where glue is applied to the relevant glue flaps for sealing to take place. Depending on the volume of job and the level of sophistication, the gluing and folding stage can range from purely manual to fully automatic.

All in all the rather low annual production does not seem to justify the adoption of a packing (cartoning) machine. Therefore a semi-automatic or even manual packing line appears more appropriate unless the line can be used for a similar product.

Question 63 – Describe the essential requirements of the packaging line for 70 ml of liquid shampoo in a glass bottle at a production rate of 50,000 bottles per week.

Fifty thousand bottles a week, assuming a five-day working week and an eight-hour day, results in an output of just twenty-one bottles per minute. This output rate is low and does not warrant investing in an automatic filling line. Therefore a simple gravity or pressure filler will be adequate.

Assuming the bottles are already available at the user's factory, a manual bottle feeding arrangement will be okay. Depending on how clean the bottles are, pre-cleaning before filling may be unnecessary since we are not dealing with an edible product. Considering the production rate per minute, closing and labeling could both be manual. However a labeling machine could be considered if a neat and symmetric labeling is expected, particularly if the bottle shape is complex. The labeling machine could also be used for other products to justify the investment.

Collating and packing for transport and distribution could be by shrink-wrapping a pre-determined number of bottles on a tray or packing the same in partitioned outer cases to avoid bottle breakage and product spillage. The ultimate choice will depend on the transportation mode, its efficiency and reliability.

Question 64 – Outline the requirements of a packaging line for canned food with a paper label. Indicate the important features of each operation.

The requirements of a packaging line for canned food with paper labels are as follows: Since food packing operation needs to be carried out in a highly hygienic environment for obvious reasons, proper handling, storage, and cleaning of empty cans are essential steps preceding the filling operation. The way the empty cans are delivered and the environment in which they are stored and handled should be such as to reduce any source of contamination such as dirt particles in cans or rust development on any part of the cans to the barest minimum. The next step is product preparation. This varies from product to product. Some product preparation may involve completely changing the constituent raw materials from which the food is made (e.g. margarine), while others may just require that the raw items be properly washed and sterilized before filling, at which point the identities of the raw foods are still preserved, such as meat, poultry (frozen chicken), etc.

After food preparation is the filling operation. The type of filling depends on whether the food is in liquid, semi-liquid, or solid form. The most essential thing here is that whether filling is automatic or not, manual handling should be minimized and all operators on the line must put on clean overalls, caps, and gloves. The ideal thing is to fill under vacuum for obvious reasons, among which are flavour preservation and expulsion of oxygen gas, which might cause oxidation and rancidity if oil or fat is part of the raw materials.

The operation that follows filling is closing or sealing, which should be hermetic, that is, be impermeable to moisture and gases. Sealing is best done on automatic double seaming machines. While processing after sealing is necessary for some canned foods, others, like margarine, no longer need be subjected to further heating and cooling since the purpose of this operation would have been taken care of at the pasteurization stage of the product preparation. Heating the canned food to a specified temperature and keeping it there for a specified period of time produces a commercially sterile product since this temperature–time regime destroys the harmful micro-organisms. This operation also reduces the amount of cooking required by the consumer later on.

The next operation is cooling, which halts any adverse effects that prolonged heating might cause to the product. Cooling also makes subsequent handling less risky to the operators. After cooling, the cans are labeled. Since seaming has been done, the possibility of any adhesive finding its way into the product is now remote. Also, the label graphics and designs cannot get damaged at this stage as a result of the previous heating and cooling, both of which are wet operations. The main advantage of this operation is that the printed label imparts some sales appeal to the otherwise plain and unattractive can. Secondly, the label carries some markings/identification about the product such as the product name, its producer's name and address, an ingredients list, nutrition claims, directions for use, weight of content, best before date, etc. Finally the cans are packed in shipping containers, usually fibreboard corrugated cases for warehousing and distribution.

Question 65 – 'Packaging may be defined as a means of ensuring the safe delivery of a product to the ultimate consumer in a sound condition at a minimum overall cost'.

Discuss the economic implications of this statement with reference to both of the following:

(a) *X-ray machinery for home and export trade*

(b) *Biscuits for the domestic market*

Packaging may be defined as a means of ensuring the safe delivery of a product to the ultimate consumer in a sound condition and at a minimum overall cost'. The economic implications of the statement with respect to (a) and (b) above are as follows:

(a) X-ray machinery for home and export trade – X-ray machines are very expensive pieces of equipment whose packaging costs are generally low compared to the equipment cost itself. While the packaging cost of some items is as much as 20 per cent of the cost of the items themselves, this is not true of the equipment under consideration. The cost of damage that will likely be incurred by using unsuitable packaging for the X-ray machinery could be staggering compared to the additional packaging/cushioning costs that would prevent the occurrence of such damage.

Secondly, X-ray machinery is not the type of product that needs to be packaged on a high-speed packing line. If any packaging machinery is engaged in the packing operation, the question of speed will be a non-issue since, unlike household consumer goods, the demand for X-ray machines in numerical sense is very low. The vital question here therefore appears to be safe delivery regardless of whether equipment is for home trade or export trade. Nevertheless one must be aware of the need to engage a more stringent packaging system for a product meant for export trade to prevent damage that may occur as result of multiple handling and uncertainty about the local transport conditions in the overseas countries. Therefore, for the type of equipment under discussion, it is safer and economically wiser to spend a few more dollars on packaging to ensure the safe delivery of equipment than trying to save on packaging and risk getting the equipment damaged. The overall packaging cost for safe delivery may not be more than 2–5 per cent of the cost of the X-ray equipment itself.

(b) Biscuit for the domestic market – The situation here contrasts very much with what we have in (a). Biscuit is a fast-moving consumer product with a limited shelf life. Biscuits are made such that the final oven-fresh product is crisp and brittle. With time the product absorbs moisture and loses its crispiness, and this is an undesirable development from the consumer viewpoint. The primary package therefore (usually a metallised plastic film) must be carefully chosen to ensure that product freshness is maintained for a reasonable period of time. Fortunately plastic films are not generally expensive.

Packing/sealing are usually performed on high-speed automatic machines. The speed of the machine, hence product output will bring down the machine overhead cost to a reasonable level.

As for the outer cases, the fiberboard (solid or corrugated) used must be strong enough to ensure that the biscuits do not get to the ultimate consumer in pieces like broken china. Overall the cost of packaging for safe delivery would be about 10–15 per cent of the cost of the product.

Question 66 – Explain briefly how you would investigate the cost of introducing a new packaging method, assessing effect of such factors as weighing and filling by machine, the standardization of packages and/or closures, and setting-up time.

When planning to introduce a new packaging method, the overall cost of the package in relation to the product cost should be at the back of our minds. The overall cost should not be more than the average for that line of business; otherwise one may have to sell at a non-competitive price.

First the packaging material must be seen as capable of protecting the product against all hazards the product is likely to encounter before getting to the ultimate consumer. Initially one should have an idea of what one wants and should provide a model, if possible, or clearly describe one's requirements to the prospective supplier. Once this is accomplished, ask for quotations and a prototype. Any quotations given will probably be as delivered to the user's factory or ex-supplier's factory. Both options should be carefully studied since delivery costs can be substantial at times. Usually, by the time all modifications to the initial model or description are made, there should not be much difference between the final cost and the initial quotation. What you get as a final price from the supplier is a function of the market forces and your knowledge of the industry.

Between asking for quotations and prototypes and making the final modification to the sample supplied, the designer should always seize the opportunity to carry out some trial runs on packaging machineries if these are already available. From such trials, ease of filling and sealing, machine speed, waste level, filling weight, and weight control should be monitored and well quantified for study and comparison with any similar operation previously or presently performed. All these factors affect the overall packaging cost. One should also avail himself of the opportunity to correct any design faults that might have been overlooked. However, it must be mentioned that the above procedure may not be possible for such packages involving the use of moulds, cylinders, and so forth since it is not an easy task to modify these items once they are fabricated.

If there will be need to pack another product or pack size on the packing line which may require the retooling of the machine to take the new pack size, the setting-up time should be built into the overall costs. In order to reduce the setting-up time, some degree of standardization may have to be established between any existing pack sizes and the new pack size.

Finally the cost of carrying out quality/quantity checks on incoming materials should also be taken into account in arriving at the overall cost of introducing the new packaging method. Of course, one should be able to achieve reasonable comparative costs with any similar existing packaging methods.

Question 67 – A package must fulfill certain essential requirements, which will influence the cost of packaging materials and the cost of the packaging operation. The product cost is an additional factor.

Discuss the packaging costs (materials and packaging operation) related to those essential requirements, using as example:

(a) 50 kg cement in multi-ply paper sack

(b) Lipstick in a carton

(c) TV receiver

The essential requirements that will influence the cost of the packaging materials and the cost of the packaging operations are

(i)   Ability of the packaging material to protect the product against all hazards throughout the product's shelf life

(ii)   Ability of the packaging material to project the image of the product

(iii)   The ease of running the packaging material on the packaging machinery at economic speed

We now look at these effects with respect to

(a) Fifty kilograms of cement in multi-ply paper sack – First the number of plies (usually two to six) and the thickness, or grammage, of each ply will be determined by the weight of the cement and the level of stacking and mode of transport/distribution envisaged. In practice it has been established that many plies of low thickness are preferred functionally to one or two thicker plies that give the same overall thickness or grammage. The most essential thing is that the bag should not burst or tear

77

in service in order to avoid cement spillage, which will result in loss and difficult handling. Also the bag must prevent caking throughout the life of the product.

As for packaging operations, methods of filling and sealing multi-wall paper sacks abound. The product output rate and economy will determine which method to adopt.

Since cement is not a beauty product, the only image the package could be called upon to project is that of ensuring that no spillage and/or caking occurs through bursts or tears throughout the handling of the product from the packaging line to the point of usage. Also the package should identify the content by the name or trademark of the manufacturer as well as the manufacturer's claim on the product.

(b) Lipstick in a carton – This is a beauty product. However, since most lipsticks are already well protected in molded plastic containers, tubes, or glass bottles, the protection offered by cartons will be very minimal. Therefore one should not consider any expensive barrier material for the carton. However, this does not mean the carton should not be made to ensure that opening, erection, and closure/re-closure will be an easy task both on the packaging line and during normal use. From a practical point of view, the function of the carton ceases during the first use in the consumer's house since most ladies do away with the carton at this stage. But before purchase, the carton must have sales appeal via an attractive graphic design.

(c) TV receiver – Most TV receivers are packed in corrugated fiberboard materials; the board grade used depends on the weight and size of the unit. This is a situation where the use of cushioning material is inevitable because of the fairly high price of the individual receivers and their fragility. The use of a high-speed packing line may not be necessary here, though this will depend on the product output rate.

For protection, top and bottom cushioning trays are usually engaged. These fit round the front, the two sides, and base or top of the cabinet. The front and back are also padded to prevent the screen and the cathode ray tube being damaged. Once finished, hardly any part of the set touches the sides of the outer case. Also the protective packaging hardly has any role in sales appeal.

Chapter 8 Packaging Economics and Related Calculations

Question 68 – List five important factors contributing to 'overall packaging cost'. Illustrate your answer by reference to the packaging of a canned food and a washing machine.

Five important factors contributing to the overall packaging cost are the

(i) Cost of raw materials

(ii) Cost of printing and conversion, if required

(iii) Cost of transporting the empty containers and insurance

(iv) Cost of storage and handling

(v) Cost of packaging machinery, labour, speed, and wastage

Most food cans are made of tinplate. The major raw materials here are low carbon mild steel and tin. A third major raw material is lacquer, but this is optional.

Printing and conversion of tinplate into cans are very expensive and constitute a major cost element in the food canning operation.

Delivery cost is also an important element, which cannot be over-looked. The mode of packing for delivery varies in scope and in cost depending on the requirements of the customer.

Since cans are rarely supplied directly to the production floor, storage and handling are almost inevitable in canning operations. And storage space, whether hired or owned by the food processor costs money. Handling, which may include sampling for quality checks and the subsequent transfer of the cans to the packing line, is also an important cost element.

Finally the use of the cans on the packaging line at whatever speed and the number of hands engaged go a long way in determining how cost effective the packaging is. In the course of running the cans, wastage is inevitable. Three major factors affect the wastage level and they are

(a) The quality of the packaging material itself

(b) The status of the packing machine

(c) The skill or competence of the machine operator or handler

Not every packaging material will have all the cost elements mentioned above to any appreciable extent. For example, the major cost elements in the packaging of a washing machine will be the raw materials and labour. A washing machine is generally packed in corrugated fibreboard material. Therefore all the stringent measures employed in respect of food cans, as illustrated above, are not necessary here. The primary concern here is the protection of the washing machine against physical damage.

Question 69 – A company produces 4.5 million laminate tubes of toothpaste per annum. If 7.92 m² of laminate is required to make 1000 tubes, calculate the amount of laminate required to meet the annual requirement of the company. If the laminate has a rated grammage of 95 g/m², how many metric tons of laminate should this be?

1000 tubes require 7.92 m² of laminate material.

Therefore, 1 tube will require (7.92/1000) or 0.00792 m² of material.

Therefore, 4,500,000 tubes will require 0.00792 × 4,500,000 = **35,640 m²** of laminate material.

Weight of laminate in metric tons will be (35,640 × 95)/1,000,000 = **3.3858 tons.**

Question 70 – Metal hidden-thread screw caps for cream jars cost a manufacturer $43.68 per 1000. He considers changing to plastic caps made of polypropylene in a linerless construction, which would cost $39.52 per 1000, plus a mould investment of $8320.00. How many caps would he have to purchase before the mould is paid off? How long would it take to amortize the mould at 4,000,000 caps per year?

Metal caps cost $43.68 per 1000.

PP caps cost $39.52 per 1000.

Switching from metal caps to PP caps results in savings of $4.16 per 1000.

Therefore savings per unit = **$0.00416.**

Number of caps to be purchased before the mould is paid off = Cost of mould divided by the amount saved per cap, that is, $8320/$0.00416 = **2,000,000 caps.**

At four million (4,000,000) caps per year, time taken to amortize the mould or pay-off period = 2,000,000/4,000,000 = **0.50 year or 6 months.**

Question 71 – A powdered milk producer uses a foil laminate material for the sachets. Each sachet material measures 100 mm × 140 mm, and the laminate grammage is 80 gsm (g/m²). If one metric ton of laminate material costing $12.00 per kg is available, and each sachet pack weighs 25 g net.

(a) *How many sachets will the producer get from each kg of laminate?*

(b) *What is the cost per sachet of laminate?*

(c) *What quantity of milk will the producer get from the avail-*

able 1000 kg of laminate, assuming zero wastage level? Is this assumption realistic? Explain.

For a grammage of 80 g/m², laminate yield = 1000/80 = 12.5 m²/kg.

Area of one unit of sachet = 100 mm × 140 mm = 0.10 m × 0.14 m = 0.014 m².

(a) Therefore number of sachets per kg = 12.5/0.014 = 892.86 = **892.**

(b) Cost per sachet of laminate = $12.00/892 = **$0.01345.**

(c) Quantity of milk packed from one metric ton of laminate = 25 g/sachet × 892 sachets/kg × 1000 kg/ton = 22,300,000 g = 22,300 kg = **22.30 tons.**

The zero wastage assumption is not realistic because in running any packaging material on an automatic line, there will always be some wastage due to material imperfection, machine status, and human factors, including the machine operator.

Question 72 – A 50 micron LDPE film, S.G. 0.925 is required on a form/fill/seal machine. The reel width is 350 mm, and the repeat length for each bag is 180 mm.

(i) What is the weight of each bag?

(ii) What is the grammage of this film?

(iii) How many bags can we get from 1 kg weight of this film, assuming zero wastage?

(i) From $\rho = M/V$, $M = \rho \times V = \rho \times L \times W \times T$ where M = mass, V = volume,

ρ = S.G., W = width, L = repeat length, and T = film thickness.

M = 0.925 × 18 × 35 × 0.005 = **2.914 g.**

(ii) The grammage of the film = 100 cm × 100 cm × 0.005 × 0.925 = **46.25 g/m².** Alternatively for monolayer films, grammage = film thickness in microns × S.G. = 50 × 0.925 = **46.25 g/m².**

(iii) Number of bags in one kg = 1000 g/2.914 g = **343 bags.**

Question 73 – *One of the most common laminates used for packing hygroscopic powdery products is 12-μ pet/adhesive/9 μ foil/adhesive/50 μ LDPE. Estimate the laminate grammage assuming the tie elements account for 5 gsm. Assume LDPE's S.G. is 0.925 and PET's is 1.36.*

PET layer grammage = 100 cm × 100 cm × 0.0012 × 1.36 = 16.32 g/m².

An alufoil of 9 μ gauge is equivalent to a grammage of 25 g/m².

LDPE layer grammage = 100 cm × 100 cm × 0.925 × 0.005 = 46.25 g/m².

Therefore total grammage = PET grammage + alufoil grammage + LDPE grammage + tie element grammage = 16.32 + 25.0 + 46.25 + 5.0 = **92.57 g/m².**

Question 74 – *One metric ton of a 400 gsm white lined chipboard is to be printed and converted into biscuit display cartons, which requires 0.0784 m² of the board to produce each carton. Allowing 10 per cent for off-cuts and other wastages, how many of such cartons can be obtained from the one metric ton of board?*

By the definition of grammage, 1 m² of the board weighs 400 g.

Therefore 0.0784 m² will weigh 0.0784 × 400g = 31.36 g, which is the weight required to produce a display carton.

Number of cartons from 1 metric ton of the board = 1,000,000/31.36 = 31887.76. Allowing for 10% wastage, this translates to 31,887.76 × 0.9 = 28,698.98 = **28,699 cartons**.

Question 75 – Corrugated cases (RSC 0201 style) measuring 295 mm × 275 mm × 175 mm are to be produced using board construction 200K/112B/200K. If the costs of kraft paper and fluting paper are US$600 and US$400 per ton, respectively, calculate: (i) the surface area of each case. (ii) The cost of material required to produce 1000 cases, assuming 5 per cent wastage. (iii) The number of cases which can be obtained from one ton of kraft paper and one ton of fluting paper, respectively. Assume a wastage level of 5 per cent for question (iii). Use 1.3 as the fluting factor for the 'B' flute.

(i) Area of board for each case = 2(L + W)(W + H), where L = length; W = width; H = height.

Area of **board** = 2(0.295 + 0.275)(0.275 + 0.175) = **0.513 m²**.

(ii) Cost of kraft liner is US$600 per ton. Liner grammage is 200 g/m². Therefore yield per ton = (1000 × 1000)/200 = 5,000 m²/ton. Therefore cost of kraft (inner + outer) per case = (0.513 × 600 × 2)/5,000 = $0.123.

Area of fluting material per case = 2(L + W)(W + H) × 1.3, where 1.3 is the fluting factor for B flute = 2(0.295 + 0.275)(0.275 + 0.175) × 1.3 = 0.667 m².

Cost of fluting material is US$400 per ton. Material grammage is 112 g/m². Hence yield per ton = (1000 × 1000)/112 = 8,928.57 m² per ton.

Therefore cost of fluting per case = $(0.667 \times 400)/8{,}928.57 = 0.02988 = \$0.03$

Total material cost per case = $0.123 + 0.030 = \$0.153$.

Material cost per 1000 cases = **$153.00**

(iii) Number of cases per ton of kraft paper = $(5{,}000/(0.513 \times 2)) \times 0.95 =$ **4,629**

Number of cases per ton of fluting paper = $(8{,}928.57 \times 0.95)/0.667$ = **12,716**

Question 76 – An alternative method for packaging an established product has been suggested. The costing information for the two alternatives is given in the table below. Making reasonable assumptions on labour costs for packaging, overheads, etc., calculate

> *(a) The saving with the new pack per unit in overall cost delivered in Moscow*

> *(b) The saving with the new pack as a percentage of the manufactured cost of the product*

State and discuss briefly the other information required before you would agree to the changeover.

EX-FACTORY COST OF PRODUCT
(materials, labour, etc.) – $124.60

	Old Pack	New Pack
Case and packing material cost	$7.40	$3.36

Packing time	14 (man minutes)	9 (man minutes)
Dimensions of pack (cm)	45 × 40 × 19	38 × 38 × 15
Weight (gross)	7.3 kg	4.9 kg
Air freight cost (c.i.f. London – Moscow)	$27.60	$20.84

Labour costs for old pack and new pack are assumed to be $3.60 and $4.00/hour respectively, thus giving ($3.60 × 14)/60 = $0.84 and ($4.0 × 9)/60 = $ 0.60 per unit of old and new packs, respectively.

Similarly overheads per unit pack are assumed to be $0.52 and $0.40 for the old and new packs, respectively.

COSTING ($)

	Old Pack	New Pack
Cost of product	124.60	124.60
Case/packing material cost	7.40	3.36
Labour cost	0.84	0.60
Overheads	0.52	0.40
Air freight cost c.i.f.(London-Moscow)	27.60	20.84
TOTAL	160.96	149.80

(a) Per cent saving with new pack is equal to

(Overall cost of old pack – Overall cost of new pack)/Overall cost of old pack × 100/1 = (160.96 − 149.80) × 100/160.96 = (11.16 × 100)/160.96 = **6.93%**

(b) Savings with new pack as per cent of manufactured costs of products, the latter being interpreted as all costs except freight.

Manufactured cost of old pack = $133.36

Manufactured cost of new pack = $128.96

% Savings = 4.40/133.36 × 100/1 = **3.3%**

Though the savings in percentage terms is not substantial, if the scale of operation is high it could translate to substantial savings in dollar terms. Before one can agree on the changeover, there must be a guarantee that the quality of the product in the new pack is at par with the quality of the product in the old pack or even better.

Questions 77 – The data below were the sample weights taken from an automatic weighing/packing machine producing a consumer good.

Weight in grams:
103.2, 107.8, 103.8, 104.4, 107.6, 103.0, 103.3, 108.2, 103.5, 107.8, 107.8, 104.6, 101.8, 103.7, 104.2, 103.4, 107.8, 104.4, 101.9, 103.2, 107.2, 104.5, 106.8, 103.6, 103.4.

(i) Calculate the mean and the standard deviation for the sample.

(ii) If 100 g is the net minimum product weight allowed for each pack, what proportion of the packs falls below the allowed minimum?

Let the individual observations be $x_1, x_2, x_3, \ldots x_i \ldots x_n$

The number of observations = n = 25 in this case

(i) Therefore the estimated mean = xbar = $\sum(xi)/n$ = 2620.9/25 = 104.836

The estimated standard deviation = s = $\sqrt{\sum(xi - xbar)^2/(n-1)}$ = 2.0738 g

(ii) The probability that pack weight will be less than 100 g = P(X < 100 g) = P(Z < (X xbar)/s = P(Z < (100 − 104.836)/2.0738); where Z = z-value of X.

= P (Z < -2.33) = 0.0099 or 0.99% ≈ 1%.

Question 78 – A free flowing granular consumer product is currently packed in a PET/FOIL/PE laminate-lined carton, and 100 g of product is declared on the carton. The carton weighs 15 g ± 2 g, and the current gross average weight is 130 g (standard deviation 2.5 g). Indicate what you would recommend to management in order that the 'fill' can be reduced and to what value.

Declared weight = 100 g. Gross average weight = 130 with S.D. 2.5 g.

Therefore, gross range weight = 130 g ± 2 × S.D. = 125 g – 135 g. Using an average carton tare weight of 15 g, net weight range becomes (125 – 15) g to (135 – 15) g = 110 g to 120 g. It means current target weight can be expressed as 115 g ± 2 × S.D.

All we are interested in is ensuring that the minimum net weight does not fall below 100 g. Assuming the S.D. remains roughly the same as

2.5 g even when we reduce 'fill' and that we do not go below 100 g, 95 per cent of the time, we can set a new target weight using the following equation:

Target weight (new) – 2 × S.D. ≥ 100 g; Substituting 2.5 g for S.D., Target weight = 105 g. Therefore, 105 g ± 2 × S.D. = 100 g to 110 g. We can reduce 'fill' by 115 g – 105 g = 10 g without any problem.

A reduction of 10 g in 'fill' will be recommended to management.

Question 79 – Write a short essay on over-packaging. Illustrate with two examples of packaging, one of which you consider to be over-packaged and one which you consider to be satisfactory. Using the example of the over-packaged product, discuss how an alternative package could be used which would be less wasteful of material but still give a satisfactory performance. Also, using as an example the package you consider satisfactory, discuss why you feel it could not be reduced without adversely affecting its performance.

Over-packaging can be said to exist if we have a situation whereby we can spend less in material and/or design than we are currently doing and still get the product delivered without a reduction in its functional performance. Examples of over-packaging abound in every society today, so much so that millions of dollars can be saved if over-packaging is avoided. Some over-packaged products are deliberate, while others are due to insufficient knowledge/data on the package itself, the product we put in it and/or the interaction between the environment and the package/product combination.

Examples of intentional over-packaging are rampant in the cosmetic/beauty industry. Here many complicated designs with multicolour printing exist, which do not enhance the functional performance of the product except its aesthetic value. The situation can hardly be helped by any outcries about over-packaging because it is a clear consequence of society's ever increasing appetite for sophistication, beauty, and class.

For example, a company was once marketing a popular brand of body cream in a single-wall Medium Impact Polystyrene (MIP) pot fitted with wadded screw-on cap, also made of MIP. Years later, the double-wall variant of the pot was adopted in place of the original single-wall pot. This is an example of not only over-packaging but also deceptive packaging because weight for weight (product), the double-wall pot looks bigger than its single-wall equivalent. The double-wall pot consumes more resin per pot than the single-wall. Also labour, equipment, and material are required to weld the inner shell of the double-wall pot to its outer shell before a complete pot is obtained since the inner and the outer shells are moulded separately. Reverting to a single-wall pot will definitely result in less material usage and labour cost with the same product performance.

An example of a satisfactory package is a tube of toothpaste. The paste is filled into collapsible aluminium tubes with the individual tube going into a display carton made from white lined duplex board. These are now collated in 12's on an over-wrapping machine before they are packed in single-wall corrugated fibreboard cases of board grade 200K/127B/200K. The cases are palletized and stacked two pallets high in the warehouse, ready for distribution.

The two major parts of the total package offering substantial physical support are the individual cartons and the outer cases. Over the years, the right choice of the individual carton board grade has been made by a careful study of the cartoning machine/carton grade performance. This leaves us only with the grade of the outer case for review. A board grade of 300K/127B/200K was initially selected for the case, and it worked without any incidents of crushing. Next was the adoption of the current grade, 200K/127B/200K, which is just a step below the initial choice. This grade also performed well. The next move at further cost reduction was a case of board grade 200K/127B/125K, also just a step below 200K/127B/200K. No sooner was this introduced than complaints of crushing began to pour in. The next and final move was to quickly revert to the use of board grade 200K/127B/200K, and this has been doing the job very well ever since.

Question 80 – A packer of canned goods pays $54.00 per 1000 cans. Experience tells the packer that for every 1000 cans delivered, he will lose 2 per cent through damage, sampling and inspection activities, and experimental work. What is his cost per 1000 cans shipped?

For every 1000 cans delivered, he loses 2%. Therefore number of cans lost out of 1000 cans delivered = 20. Number of good/usable cans = 980. Therefore, cost per 1000 cans shipped will be ($54 × 1000)/980 = **$55.10.**

Chapter 9 Food, Drink, Pharmaceutical, and Cosmetic Packaging

Question 81 – What are the common mechanisms of food spoilage? How can packaging assist in protecting food against each of these mechanisms?

The common mechanisms of food spoilage are

(a) Microbial – The spoilage here is due to the activities of bacteria and fungal moulds. These organisms flourish under warm conditions rich in oxygen and moisture. The way packaging can assist in protecting food against spoilage is to ensure that, once the food is sterile, the packaging material is such as to prevent the entry of these organisms into the containers. The packaging material itself should be free of the organisms and be good barriers to gases and water vapor.

(b) Oxidation – Spoilage here is due to oxygen pick–up, which results in fats/oils breakdown due to oxidative rancidity.

(c) Moisture pickup – Most dry and granulated products easily pick up moisture with attendant quality deterioration. Packaging must have enough barrier properties to prevent this.

(d) Moisture loss – This is basically the reverse of (c). In order for packaging to prevent spoilage in (b), (c), and (d), a material that offers reasonable resistance to oxygen, water, and water vapor, as the case may be, is required. The basic requirements for individual food items vary and will depend on the nature of food, the environment, and the shelf-life required, among other things.

(e) Light – Light is known to accelerate some biological and chemical processes that have adverse effects on food. If product is known to be sensitive to light, the best solution is to use opaque or coloured packaging materials, provided these do not create other problems.

(f) Mechanical damage – No product or ware, no matter how rigid, enjoys absolute immunity against mechanical damage. The solution for this calls for a compromise between the unit pack packaging and the shipping container packaging with due regard to storage and distribution handling facilities.

(g) Odor and flavor pickup – Usually the measures taken in (b), (c), and (d) will take care of odour/flavour pickup.

Question 82 – Describe two common methods of food preservation. What demands does each of these make on packaging materials?

The first method of food preservation is the heating process. This is referred to variously as thermal processing, cooking, or retorting. The idea is to subject a packaged product to heat and hold it there for a pre-determined period of time. The processing temperatures and holding time vary from product to product. The objective in each case is to ensure that those micro-organisms, especially Clostridium botulinum, that cause food spoilage are reduced to the minimum possible without destroying the nutritive values of the food itself.

The demands heat processing make on food packaging are many. First the packaging material must be able to withstand heat processing. For example glass containers must not crack due to thermal shock. Some plastic materials (e.g. LDPE) have been found unsuitable for food packaging because they have rather low softening points and cannot stand most conventional heat processing methods. Also, in order to ensure good heat penetration of the food during heat processing, the wall thickness and the actual shape of the packaging container have to be properly controlled. Otherwise food products in certain regions of the container will not benefit from the heat treatment while those close to the container walls may be over-treated. Also packaging must be impervious to the process fluid.

The second method of food preservation is by low temperature process. This is more or less the reverse of the heating process. This method is based on the premise that both chemical and microbiological processes are slowed down as the temperature is reduced. The disadvantage of this method is that the organisms are only reduced to a state of inaction. As soon as the temperature rises, the organisms resuscitate.

The demand on packaging by this preservation method is that material must be compatible with the product at the low temperatures used for storage. Also the packaging material must not break down (e.g. delamination in laminate structures).

Question 83 – Discuss the advantages and disadvantages of glass bottles and alternative packs for liquid milk

The inertness and barrier properties, clarity, rigidity, and heat resistance of glass make it very suitable for milk packing. The inertness and barrier properties make the glass bottle preserve the content due to inaccessibility of any micro-organisms once the product and the container are sterile at the filling stage and the closure is airtight.
The clarity property confers sales appeal on the milk while the container rigidity makes repeated use of the container a very cost-effective package. Cleaning of the container before reuse is made possible by the container ability to withstand heat treatment.

Vitamin C in milk can be destroyed by heat, light, and oxygen while its vitamin B can be destroyed by heat and light. Most glass bottles for milk are white and transparent. Therefore the effect of light on vitamins B and C could cause milk deterioration, and this is a disadvantage. Fortunately the time required for light to effect damage is generally longer than the time over which most milk products are kept. The use of coloured glass bottles does not help the situation because colored bottles absorb heat more than flint bottles, and excessive heat is one of the causes of milk deterioration.

Alternative packaging materials normally used for milk are wax or polythene lined cartons, plastic sachets, and plastic bottles. Since these are non-returnable (one trip) containers, they pose the problem of disposability in the large quantities dictated by milk sales. Nevertheless these alternatives are cheaper unit containers than glass. The alternative packs are not totally as impermeable to gases, vapor, etc., as glass. Hence the possibility of early product deterioration is there. Finally none of the alternative packs are as rigid as glass. Therefore they will all need stronger and more expensive shipping containers for efficient and safe distribution. Modern high barrier plastic bottles are now available through co-extrusion technology, which gives high barrier properties almost at par with glass bottles. These can serve as alternatives to glass in the nearest future.

Question 84 – Discuss the factors, which influence the shelf-life of fresh fruits and vegetables. Show, with examples, how shelf-life can be extended by good packaging practice.

One notable feature of fresh fruits and vegetables is that even after they have been harvested they are still living organisms and normal metabolism must be maintained.

This means the absorption of oxygen (respiration), which breaks down carbohydrates to give water and carbon dioxide, continues. If oxygen is restricted, a different chemical reaction comes into play and small quantities of alcohol are produced resulting in off-odours and off-flavours together with a break down of plant cells. This process is known as anaerobic decay because it takes place in the absence of oxygen. On the other hand, spoilage which takes place in the presence of air is called aerobic decay.

The above has revealed that packaging fresh fruits and vegetables in airtight material could result in undesired effects. At the same time, leaving the products unpackaged leads to excessive moisture loss, with consequent wilting of the vegetables. All these affect the shelf-life of the produce with attendant early deterioration. To get over the above problems, a compromise in the choice of packaging material is inevitable.

The chosen packaging material must be such as to allow the ingress of oxygen for respiration and the escape of carbon dioxide and water released by the produce. Unfortunately most vegetables respire at a fairly rapid rate, and only porous or perforated wrappings let oxygen and carbon dioxide through quickly to preserve product freshness. The usual solution is to pack or bag the wares in a near impervious wrapping having a few holes about 5 mm in diameter, or to leave the pack only partially sealed. LDPE film is commonly used for the pre-packaging of a wide variety of produce. And because of its low moisture vapor permeability, it needs to be perforated to avoid condensation; otherwise, the resulting environment within the package will be a potential colony for the growth of bacteria and moulds. Where respiration rates are low, it is possible to use un-perforated films with a high permeability to oxygen and moisture vapor such as cellulose acetate and regenerated cellulose.

Soft fruits (e.g. strawberry, blackcurrant) are extremely susceptible to mechanical damage; hence they need to be packaged in semi-rigid containers, such as moulded pulp or plastics, punnets, etc.

As with other biological and chemical changes, respiration and its associated activities are slowed down by storage of produce in low temperature environments. The fact that ripening is also retarded by low temperature with attendant blackening of the produce, low temperature storage is not advised. Hence a compromise temperature is often necessary.

Question 85 – Write a short account of developments in the packaging of carbonated beverages.

The main packaging requirements for all carbonated beverages is that it should retain the carbon dioxide gas, without which the drink would taste flat or stale. Another requirement is that the package must be able to withstand the internal pressure generated. The first requirement calls for a material that possesses high gas barrier properties while the second requirement calls for a material with good strength and rigidity.

Glass bottles were the original choice, and these are still widely used with crimped on crown cork closure. Next to the glass bottles are cans (tinplate and aluminium) made by the 'drawn and wall ironed' process, and most of these are fitted with the 'tab pull' easy-open ends or aluminium 'ring pull' easy open ends.

Other packs used for carbonated beverages include the Rigello pak and Merolite. The Rigello pak consists of a vacuum formed PVC body with a convex base to which is welded a cone shaped shoulder and neck. The plastic body is enclosed in a spirally wound fiberboard tube with an outer surface of gravure printed aluminium foil. A specially designed injection molded polythene cap completes the pack. Merolite is the best-known truly flexible pack for carbonated beverages. Merolite is made of a laminated material incorporating polyester film for strength and polyvinylidene chloride (PVDC) for barrier.

Other plastic options either being tested or being developed are blow-moulded PVC bottles and modified acrylic bottles. Even in the traditional glass bottles, changes are being made to overcome some of its shortcomings. An example is the development of lightweight one-trip bottles with plastic lined screw caps or metal twist-off crowns.

The latest pack for carbonated beverages is the emergence of blown PET bottles, which is now well known and universally accepted as a suitable alternative to glass bottles. This material has the advantage of a light weight, and consequently, a low cost compared to glass.

Question 86 – What are the factors involved in the choice of packaging material for bread? Show how the various requirements are fulfilled by the packaging already used for this purpose and discuss any shortcomings.

Bread can be regarded as belonging to the class of products referred to as moist wares. These goods lose moisture when exposed to the atmosphere and become dry and stale, so a barrier wrapping is required to prevent this loss. But some caution needs to be exercised in packing bread in an air/moisture-tight material. First oven-fresh bread ready for packaging is still fairly hot, or at least warm. Secondly, since bread is a moist ware, it has the tendency to lose moisture to its immediate environment. All these constitute an environment very conducive to mould growth. We, therefore, need a material which will give a balance between excessive moisture loss and no moisture loss at all.

Bread is now wrapped mostly in LDPE films and waxed papers. The use of LDPE films without pinholes appears dangerous in that the permeability of 25 μ LDPE film is just around 18 g/m²/day. The permeabilities of higher gauge films are somehow lower than this. The only saving grace here is to allow the loaves to cool sufficiently before wrapping. Another point in favour of film wrapping is that bread has such a high turnover that a very short shelf-life of only a few days is all that is required, and most microbiological activities would not have spoiled the bread before consumption.

The use of wax papers appears satisfactory since a 55/70 gsm waxed paper has a permeability of about 600 g/m²/day, a figure that makes one feel a lot more comfortable than the LDPEs. The only disadvantage here is the rather opaque nature of waxed paper which makes product visibility impossible. The product will dry out faster in waxed paper and become stale.

Question 87 – A cosmetic hand cream deteriorates rapidly in contact with air. Suggest ways of packaging it and devise a program for testing their suitability, giving reasons.

The fact that the cream deteriorates rapidly in air suggests that the cream is sensitive to oxygen. Therefore, in order to prevent rapid deterioration, the cream needs to be packed in a material that is a good barrier to oxygen. The closure should also be nearly airtight.

Many packaging media offer an almost total barrier to oxygen/gases. Examples are glass, tinplate, and aluminium containers. Other media offer reasonable, but not total barrier to oxygen/gases. This latter group includes polypropylene, PVC, Polystyrene, and other plastic containers.

Those media that are 100 per cent impermeable to oxygen may be rather expensive compared to the latter group that offer only a partial barrier. Therefore all options should be considered. After all, it may turn out that we don't need a total barrier medium. The final choice will depend on how stable or sensitive to air the cream is. For not too sensitive cream, wide-mouthed glass or plastic jars with good screw-on closures may be adequate. However, for very sensitive cream with fairly long shelf-lives, collapsible aluminium tubes may be inevitable.

For testing the suitability of packaging, the following program may be adopted depending on the desired shelf-life of the product:

(a) Carefully pack the freshly produced cream in the proposed container and close securely.

(b) Store a fairly large number of the filled/closed containers in an environment in which the cream is going to be retailed. If no conditioning equipment is available, the packs could be displayed outside to simulate the normal distribution and re-tailing environment.

(c) A control pack may be necessary if this is possible; otherwise, a subjective assessment may be adopted.

(d) Assess the quality of the product as freshly produced in (a).

(e) At intervals of about two weeks, repeat the quality assessment of the product as contained in the proposed packaging material. At each testing stage packs previously opened, tested, and re-closed and those, which have not been opened during any of the previous assessments, may need to be checked to establish the effect of previous openings/re-closures on product quality. The findings here may help in choosing the type of container and a dispensing method that will limit air contact with product during use.

Question 88 – Assessment of a product is an essential prerequisite for good packaging. Discuss this with reference to the packaging of (a) a liquid aftershave lotion and (b) hygroscopic pharmaceutical tablets.

Any product meant for public consumption must of necessity meet some quality requirements. Hence product quality/property assessment helps a great deal in selecting good packaging.

(a) A liquid aftershave lotion – First, the formulation of the lotion will have to be tested and confirmed as suitable for the function it will be called upon to perform. The lotion must be gentle to the skin and not cause damage to the skin. Having ascertained product quality, the packaging must be able to protect the lotion against any early deterioration. I say early because no product/packaging marriage exists that prevents deterioration indefinitely. What is expected of packaging is that it should keep the product in good condition over a period considered to be a reasonable shelf-life for the product.

In the assessment of the product, particular attention should be paid to those factors that are likely to cause product deterioration. Examples are the effects of air/gases, moisture, heat, or perfume loss on the lotion. Once these factors are identified, and quantified where necessary, then the packaging designer should be in a position to come up with a packaging container that will meet these requirements. Finally, keepability tests such as those outlined above in question eighty-seven will have to be carried out in order to ensure that the adopted packaging is satisfactory.

(b) Hygroscopic pharmaceutical tablets – Since these tablets are hygroscopic, one ought to be able to determine its critical moisture content, this being the moisture content high enough to render the product unmarketable. An estimate of the critical moisture content coupled with the expected shelf life and the humidity of the environment should aid the packaging engineer in arriving at a packaging material which will provide enough moisture barriers for the products. Most pharmaceutical products are packed in glass containers, which are known to have good moisture barrier properties provided the closure is equally good. Having gone this far, the usual product keepability assessment as outlined in question eighty-seven above should be conducted. In recent times, moulded plastic containers in various colours have been used for hygroscopic pharmaceutical tablets. HDPE and PET are the most common materials for the plastic containers. They are all, almost without exception, fitted with screw on plastic caps and induction inner-seals across the container rims (opening) to provide not only a hermetical seal, but also tamper evidence.

Question 89 – Discuss the advantages and disadvantages of each of the following alternative packs for a cosmetic cream: PVC or aluminium collapsible tubes.

The advantages of a PVC tube for cosmetic cream are that it is corrosion resistant, leak-proof, and unbreakable. It can be made transparent or made opaque by the addition of a pigment or colourant. It can be decorated by a variety of decorating/printing processes. It regains its original shape by recoiling after use. Finally it is cheaper than aluminium tubes. However, PVC tubes have the following disadvantages. The recoil action referred to above leads to air intake, which accelerates drying out of cream with consequent early product deterioration. Also processing often needs the addition of plasticisers, some of which may not be compatible with the cosmetic cream. Finally PVC requires the use of heat energy to effect sealing.

Aluminium collapsible tubes have the advantages of being a total barrier to air, moisture, light, and essential oils. With good closure and crimping, one is sure of a perfect barrier. Also when squeezed during use, the tube maintains complete deformation, hence little or no air intake occurs, unlike with the PVC tube. The dispensing nozzle is very narrow, resulting in economic product use. Finally the sealing of the base is done mechanically by folding and crimping and therefore does not require any investment on energy. The disadvantages are that aluminium tubes could be corrosive in very alkaline environments, hence internal lacquering may be necessary. This makes an already expensive tube even more expensive.

Question 90 – *What are the main hazards against which pharmaceutical products must be protected? Suggest packages which would give adequate protection against the hazards in the case of (a) a liquid medicine, (b) a tablet, and (c) an ointment.*

What other factors have to be taken into account in addition to protection? Give an example of each.

The followings are the main hazards against which pharmaceutical products must be protected:

(1) Moisture pickup/loss

(2) Volatility

(3) Heat

(4) Light

(5) Oxygen and carbon dioxide pickup

(6) Contamination by extraneous substances (e.g. metals)

(7) Adulteration

(a) A liquid medicine – This could be affected by all the hazards above if not well protected. A coloured glass container with good closure and adequate ullage will take care of all the hazards listed above.

(b) A tablet – The packaging proposed in (a) will also be adequate except that no substantial ullage is necessary. Where there is ullage, a stuffing material must be used to keep the tablets in place so that rattling and crumbling of tablets are prevented.

(c) An Ointment – Glass, plastics, or metal containers may be suitable. Aluminium collapsible tubes are also a suitable packaging material for an ointment.

Other factors which have to be taken into account in addition to protection are

(i) Function – This is essential particularly when it comes to dispensing or use. For example, some ointments may need special application to certain areas of the body direct from the container. In such cases some dispensers may be built into the package.

(ii) Appearance/sales appeal – This is not very important as far as pharmaceutical products are concerned. This is not to say surface design should not be presentable enough.

(iii) Cost – Cost will be dictated mostly by the degree of protection the packaging is called upon to offer. For example, a product which requires the use of glass or tinplate container will cost more in terms of packaging than the one that calls for the use of paper or other flexible packaging containers.

(iv) Child safety – This can be addressed by the use of a Child-Resistance Closure (CRC) on the containers.

(v) Tamper evidence – This can be addressed by use of a tamper evidence closure.

Chapter 10 Package Testing, Quality Assurance, and Specification Writing

Question 91 – State the three main reasons for making tests. How can these reasons affect the type of tests used?

The three main reasons for carrying out tests are

(i) To predict performance

(ii) To control quality

(iii) To obtain information about the strengths and weaknesses of a particular pack when exposed to an individual hazard

The above reasons can affect the type of tests used in a number of ways. If the purpose of performing a test is to predict performance, then first of all, one would like to know the performance of interest (e.g. resistance to drop, ability of package to withstand reasonable stacking height over a period of time) before knowing what tests would be most relevant. Some performance tests are carried out on total packages (including the contents) while others are carried out on the empty packages, that is, without the contents.

It appears that tests meant to control quality (e.g. gauge, grammage, dimensions, print quality) normally follow those for predicting performance (e.g. stacking, transit, drop tests), and the former are usually routine in nature while the latter end once adoption of the material under test is made. This seems logical in that a specification for the supply of any material is approved as a result of preliminary performance tests. Having set up the specification, all that is required is the ability of user/supplier to ensure compliance, hence quality control tests. Ironically the quality control tests need not be the same as the performance tests used in establishing the specification. For example, while stacking trial, drop test, and compression tests might have been used as a basis of establishing the right grade of a fibreboard case, it may no longer be necessary to use the same tests for controlling the quality of subsequent incoming cases. Determination of component grammages of the board from which the cases are made may be all that is required for ascertaining compliance with specifications.

Question 92 – Why is conditioning so essential before making tests on many materials and containers?

Properties of certain materials are a function of the environmental conditions in which they find themselves. For example, the thickness, or caliper, of a piece of chipboard varies with the humidity of its environment. The same is true of the moisture contents of some materials, that is, finished products, raw materials, and packaging materials alike.

Therefore, in order to get over this problem, some testing conditions are usually agreed upon, which are regarded as standard testing conditions. Equipments/instruments are now available to simulate these agreed conditions. For instance, most paper and board materials are now conditioned for twenty-four hours in humidity cupboards maintained at a temperature of 23°C and relative humidity of 50 per cent before tests are performed on them. The advantage of conditioning is that analysts in temperate regions, like Europe, and their colleagues in tropical regions, like Africa, who have conditioning equipment at their disposal, can carry out tests on identical materials under the same conditions and obtain very close results. This is the only arrangement whereby a buyer in one country could ascertain that his supplier in another country of different climatic conditions has produced his material to an agreed specification. The ability to subject materials to conditioning at a specified temperature and humidity facilitates international trade without rancour among trade partners.

Question 93 – For what reasons should field trials be made?

Field trials should be made because all laboratory experiments, tests, and simulations are mere imitations of the real thing and can never represent the exact conditions which packages or packaging materials undergo during actual field use. In other words any laboratory findings do not give the actual results most packages undergo in the real distribution network.

While laboratory tests are designed to provide rapid and reproducible quantitative results in which the environment has been simplified, field trials go at least a step further by providing information on

(i) The nature of damage which occurs in practice with a given package

(ii) The suitability of a pack for a specific distribution system

(iii) Correlation of field performance with the laboratory tests

In designing field trials, all factors that may affect package performance in any way (e.g. transport system, climatic hazards) and handling/storage facilities should be taken into consideration. Hence for a package design that is going to be put into use for a fairly long period of time and for a mass produced article that will enjoy wide distribution, field trials give more meaningful and practical results than any other means of package design assessment.

Question 94 – Construct a chart showing the principal mechanical transit hazards likely to occur during the distribution by road in the UK of a fibreboard case containing domestic electrical equipment weighing 15 kg from the manufacturer via a wholesaler to a retailer. Describe a laboratory test sequence, which you would use to assess the suitability of the fibreboard case. State the criteria used in assessing the results and discuss the correlation between the test and the journey.

In the following chart it is assumed no warehousing will take place at the manufacturer's end *and* the experience at the wholesaler's warehouse will be the same as at the manufacturer's end.

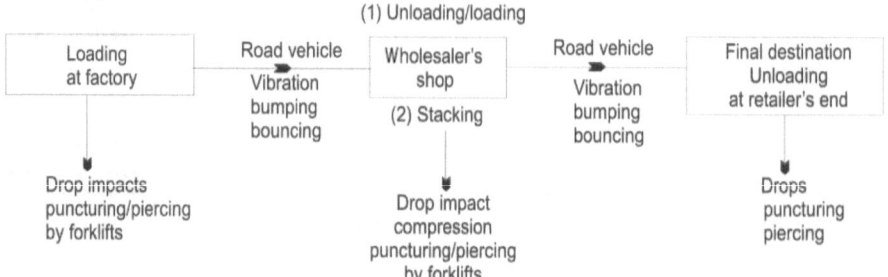

The major mechanical transit hazards likely to occur are drops, vibration, compression, and puncturing/piercing.

(a) Drop impacts can be investigated by the use of table drop testers using various likely drop heights and allowing the package to hit the platform at points where damages are likely to be sustained.

(b) Vibration could be investigated by using the vibrating table tester or an electromagnetic accelerometer can be used for visual display of the amount of vibration the package is subjected to.

(c) Compression could be investigated by the use of compression tester to determine the failing load of the package.

The amount of puncturing and piercing experienced is more a function of how disciplined the warehouse crew are. A very careful forklift driver may not punch or pierce a single case during an eight-hour working shift while a rather careless one may render useless ten cases during the same working period.

The criteria used in assessing the results and the correlation between the test and the journey are matters that depend more on the experience of the packager than anything else. For example, an empty fibreboard case

110

which appears satisfactory by a compression tester experiment may behave differently when filled with articles it is designed to contain and sent out on a rough journey. This is so because some unit containers give little or no reinforcement to the outer case containing them (e.g. flexibles), while others give substantial strengthening effect to the outer case (e.g. metal cans, glass bottles).

Question 95 – As packaging manager, you are responsible for the export of tablets of an expensive toilet soap from the UK to South Africa. The tablets are individually wrapped and then placed twenty-four in a solid board sales box, and twelve of these boxes are then contained in a solid board shipping case. Transport is expected to be in batches of about half a ton and will be by road and sea transport. Outline the tests you would have carried out to ensure the safe arrival of your product.

In this question it is assumed that the wrapping of the individual tablets is satisfactory. The tests will aim at establishing the suitability of the combined sales box/shipping case during transit. It will also be assumed that the dimensions of both the sales box and the shipping case have been suitably chosen. This is essential because either under or over-specified dimensions will have adverse consequences on the containers of this nature whatever the quality of the materials of which the containers are made.

We know the tablets will be transported by road and by sea. Therefore the major mechanical hazards likely to be encountered are drop impacts, vibration, bouncing, and bumping, pitching, and rolling on the ship. It will also be assumed that no warehousing of the packages throughout the journey to South Africa will take place, nor will any climatic conditions be considered a threat to the product.

Soap tablets are not fragile, and as a result, some of the likely transport hazards will have little or no effect on them. For example, the level of vibration, pitching, and rolling normally encountered in transit is not expected to affect the packages badly. This leaves us with two main transport hazards to be considered – drop test because packages may drop during loading and unloading and compression tests because incidents of stacking packages too high in the ship decks and/or loading some other heavy goods/objects on the packages cannot be ruled out.

Drop test – Since soap tablets are not fragile, a few packed cases subjected to some drops at pre-determined heights will be adequate. A visual assessment of deformation or damage suffered by the cases and/or the tablets will give an indication of the suitability of the package. Although crushing or deformation has no adverse effect on soap tablet washing/functional performance, for sales appeal reasons, it is essential the cases and the tablets get to the final destination/consumers un-dented.

Compression test – With some idea of the stacking height to which the packages are likely to be subjected, one should arrange to stack an equivalent number of cases for a period not less than the length of the journey to determine if the cases will crush. While a compression tester could aid this investigation, the problem of reliable correlation remains. Hence the actual stacking trial is preferred if practicable.

Question 96 – What are the salient points of each of the following tests, and what are their relevance in the assessment and performance of the container concerned?

    (a) Puncture test on solid fibreboard

    (b) Dish test for water vapour permeability of barrier

       materials

    (c) Stiffness test on folding boxboard

    (d) Stress crack test on low density polyethylene for bottles

    (a) Puncture test on solid fibreboard – A puncture test will give the energy necessary for penetrating the board by a specified pointed head. The test piece is clamped between two horizontal plates, one of which has a circular opening and the other a triangular opening. A pointed head on a pendulum penetrates the test piece, and the energy necessary to force the head through the test piece is measured. The relevance of this test in the assessment and performance of a container is that during handling of containers, objects sharp or blunt, such as forklift blades, do accidentally run into the cases on any of the sides, and the puncture energy of the board is a measure of its ability to resist or withstand such accidental punches.

(b) Dish test for water vapour permeability of barrier materials – This test measures the amount of water vapour passing through a given area of a barrier material with time. Interest in barrier materials, be they single films or laminates, lies in their ability to contain or reduce substantially the amount of water vapour or gases passing through them at a given time. The relevance of the test is that certain products packed or wrapped in these barrier materials are sensitive to water vapour and/or certain gases (e.g. oxygen); hence the amount of such gases/vapour allowed into the product must be controlled by the use of suitable barrier materials. Also some other products have the tendency to lose their moisture or some volatile constituents like perfumes, preservatives or flavouring. Such products also need good barrier materials for adequate protection that will see them through a reasonable shelf life.

(c) Stiffness test on folding boxboard – This could be measured by the use of a Taber stiffness instrument, usually in both the machine and the cross directions. The performance on the packing machine of cartons made from the board depends on the stiffness of the board. Boards of low stiffness, that is, weak boards, are not suitable for automatic packing machines. Therefore one should be able to assess the level of stiffness required for a board so that efficient packing on automatic lines can be guaranteed.

Also after packing and sealing the individual cartons on the line, the cartons are packed into outer cases, which are sealed by adhesive application, stitched, or taped before dispatching same to the warehouse for stacking. Usually the strength (stiffness) of the individual packets in the outer case reinforces that of the case, and this helps the stacks to stand better and longer in storage.

(d) Stress crack test on LDPE for bottles – This test gives a measure of resistance of LDPE to certain polar liquids or vapour under conditions of stress. When LDPE is stressed multi-axially while in contact with such polar liquids, cracking may occur. The condition for cracking is that the stress and the stress-cracking agents be present simultaneously. Typical stress-cracking agents are detergents, certain essential oils, and nitrobenzene. This phenomenon referred to as environmental stress cracking (ESC) limits the use of LDPE bottles for carrying certain organic chemicals. However, the use of high molecular weight grades reduces the tendency of LDPE to stress crack.

Question 97 – Write approximately 200 words on each of the following:

> *(a) Migration hazards from plastic films*
>
> *(b) Wet-strength papers*
>
> *(c) Rub testing of printed surface*

(a) Migration refers to the transfer of materials from a package to the package contents. The migrants can be residual monomers, catalysts, antioxidants, and other polymer additives. Some of these materials migrate or leach out of containers into the surface and/or into the body of the products they contain. Some of the products are food items or pharmaceutical preparations.

Since some of the monomers are known to be carcinogenic, the need arises to control their levels in the products. Examples of these hazardous monomers are polyvinyl chloride and acrylonitrile monomers, which have been found to be carcinogenic. Since these plastic materials are now increasingly used in the food and pharmaceutical packaging industries, many governments the world over are enacting laws limiting the level of these monomers in foods. In addition to establishing acceptable (safe) levels in edible products, methods of determining the actual levels in food items have been satisfactorily established so that compliance by both packaging and the food industries can be enforced.

(b) A wet-strength paper is a paper which remains strong when fully saturated with water. It is a paper specially treated with resins so that it does not disintegrate when wet or lose all its strength under such conditions. Such papers must retain not less than 30 per cent of their strength when measured (dry and wet) by any standard method, such as the bursting strength test.

The interest in wet-strength, paper-based packaging has to do with the use to which the material is put. For example, wet-strength papers/wrappers are required for moist foods like margarine/butter where vegetable parchment is used as wrappers. Also wet-strength papers are useful for outside packaging where they are required to withstand adverse weather conditions without compromising the quality of the products they contain. Examples are multi-ply paper bags/sacks for cement, fertilizer, and other farm inputs. These multi-wall paper sacks have now largely been replaced with plastic sacks.

(c) In printing, it is expected that the printed jobs or surfaces will always be subjected to everyday handling. A good quality print should be able to withstand such everyday handling without the ink lifting from the surface. Printed surfaces should therefore be subjected to rub proofness tests after drying to ensure that the ink is reasonably stable.

The usual test involves the use of Patra rub proofness tester or similar instruments. The test essentially calls for subjecting a cut sample of the printed surface to about 100 rubs under a standard pressure of 7 kpa. The sample is then examined for smudging or ink transfer.

Question 98 – What is a standard, and what is rationalisation? What are three important economic advantages that can be achieved through standardisation in the mass production industries? Where can standards be of use in packages which have the main function of selling the product? Illustrate your answer by examples.

A standard may be defined as an agreement among all the interests concerned on a product, service, procedure, practice, manufacturing process, or level of performance. A written record of such an agreement is known as a standard. This may include drawings, figures, and tolerances.

Rationalization is simply a reduction of unnecessary variety.

Three important economic advantages that can be achieved through standardization in the mass production industries are

(i) Planning of production becomes possible from raw material to finished article.

(ii) Standardization assists in the elimination of waste materials.

(iii) Standardization helps to improve output and quality control by use of standard test methods.

Without standardization there could not have been any mass production of goods, a brainchild of industrialization. Without standards, the use of packaging materials on fast-moving packing machines would not have been possible, because a supplier would have found himself in a difficult position producing thousands or millions of materials to a buyer's specific requirements without serious deviation. Also buying the material from different sources without standards would have meant getting as many different materials as the number of suppliers. Without standards on quality, a measure of any important attribute would be difficult to assess. For instance, how would we assess the resistance of inks and varnishes to rubbing if no standard were available? The same is true of shades of colour used and the variations allowed.

Outside the production floor, the absence of a standard could render the finished product worthless. Uncontrolled colour variation in packaging materials, for example, could confuse the consumer and make her decide not to buy because she is not sure if she is buying the same product she bought last time.

Functionally, without a standard the packaging material may not perform one of its major functions of protecting the contents. Suppose a moisture sensitive product is packed in a material without adequate moisture barrier properties, or the packaging material is not strong enough to stand the storage and transport hazards. The consequences of a lack of standards could be very grave.

Question 99 – (a) Name three basic Standards.

(b) State the five different kinds of applied Standards which are useful in the packaging field.

The three basic standards are

(i) Terminology

(ii) Definition of units and symbols

(iii) Methods of measurement

The five different kinds of applied standards that are useful in the packaging field are

(i) Dimensional standards

(ii) Performance or quality standards

(iii) Standard methods of testing

(iv) Standard technical terms and symbols

(v) Standard codes of practice

Question 100 – State the four levels of Standardisation and explain why their production becomes more difficult as one goes from lower level to the top level.

The four levels of standardization are

(i) Company standards – This is the lowest level of standard. It involves just a company or an organization, and hence it is easy to reach a decision on the standard to be adopted.

(ii) Industry standards – This is the second level of standardization. This involves companies operating in the same industry, such as the auto industry. Standards here may be produced by trade associations, etc. It becomes a little more difficult here to establish a standard because conflicting interests arise among the companies concerned, and these need to be resolved and an agreement accepted by all.

(iii) National standards – This is the third level of standardization. Here companies in diverse industries are involved, and this makes setting up standards difficult because of goal conflicts. Usually a compromise and/or sacrifices are inevitable; otherwise, it may be difficult, if not impossible, to arrive at a common standard.

(iv) International standards – This is the topmost level of standardization. Many factors (e.g. economic, political, cultural, and technological) will be called into play because nations with various political ideologies, cultural backgrounds, and technological development are involved. Let us consider the technological development factor as an example. In setting a standard for a group represented by countries from the developed and developing regions of the world, a lot of difficulties arise. The developing countries may see the standard as being too high for them to meet due to their low level of technology, while the developed countries like the USA and Britain may not like to reduce the high standards they are already used to because they want an international standard.

We can go on and on with examples which could make setting up international standards an uphill task. No wonder international agreements have never been easy to finalise. It is therefore not unusual to take five years or more before many international agreements are concluded.

Question 101 – What is a packaging specification? Why are packaging specifications necessary?

A packaging specification is an accurate and detailed description containing information about all the necessary facts, properties, and special features of a packaging material to facilitate unambiguous communication between the manufacturer (supplier) and the user (buyer).

There are five main reasons why written packaging specifications are mandatory:

(1) To ensure that the package is compatible with the requirements of the product to be packed, the eventual packing machinery used, the expected stresses/hazards of the particular mode of transport selected, and the needs of the final consumer.

(2) To avoid misunderstandings regarding both the technical and commercial details in the transaction.

(3) To serve as a basis for the settlement of eventual claims, if any should arise.

(4) To enable the buyer to search for alternative suppliers and facilitate a comparison of the different offers received.

(5) To provide the user staff with a basis for accepting or rejecting incoming packages and components.

Writing a packaging specification requires a full understanding of what performance factors are critical (e.g. to the protection of the product, to good machinability, and ability to withstand the rigours of warehousing and distribution).

Great care must be taken to establish the correct tolerance levels for every critical performance factor. Too broad/generous a tolerance can cause machine and/or aesthetic problems, while, on the other hand, too tight a tolerance may reduce the number of potential suppliers and can significantly increase costs.

Note that all aspects of a packaging specification must be mutually agreed on between the supplier and the user and on no account must any area of the specification be imposed on the supplier and vice versa.

Finally the following issues must be borne in mind and must serve as a guide in managing packaging specifications all the time.

(a) Is there a consistent corporate specification format?

(b) Who is responsible for writing the specifications? Who is responsible for authorizing?

(c) How are the specifications written?

(d) How are they issued? Who gets a copy, and where they are kept?

(e) How is a specification revised, and how is a specification withdrawn?

(f) How is a specification implemented?

(g) What is the course of action when events occur that are outside the specification?

Question 102 – Write a packaging specification for a folding carton for packing 200g of a detergent powder.

Company Name: XYZ Company Ltd
Packaging Item: Display Box for 200 g of Detergent Powder
Ref. Number: xxxxxxxxxxxxx
Date: 12 April 2008

Scope: This specification describes the display carton/box to contain 200 g of detergent powder produced for wide distribution in a local market.

Construction: The carton shall be made of 350 g/m² (gsm) white lined chipboard (WLC). Blanks for the carton shall be dimensioned as per the drawing specification number xxxxxxxxxx. The grain direction of the board (blank) shall be horizontal to the print as shown on the drawing specification. The board substance (grammage) tolerance shall be +/- 5%, while tolerance shall be +/-0.5 mm on any of the basic dimensions.

Dimensions (Internal): Length = 118 mm; Width = 36 mm; Height = 180 mm.

Printing: Printing shall be by offset lithography or by the gravure process. It must comply with the agreed artwork and colour guides. The print must be light stable while the varnish must be scuff resistant, that is, the print must not lose colour or fade on prolonged exposure to UV light, nor must it abrade when rubbed against itself or any similar surface, plain or printed. Print must be stable to water, soap jelly, and 5 per cent alkaline solution. Finally the inks as well as the board on which the printing is done must be odour-free.

Gluing: There must be evidence of fibre tear on every glued area when pulled apart. There must be no undergluing, which may result in glue failure or overgluing, which may make carton erection on the packing/cartoning machine difficult or impossible. The glue must be odour-free.

Packing and Delivery: Cartons in flat/collapsible form must be wrapped up in bundles of fifty or any convenient number and twenty of such bundles be arranged edge-on in corrugated boxes or in lined wooden boxes. No visible damage must be evident on any part of the cartons supplied as described above. Every consignment must be protected in transit against any inclement weather, such as rain or dust as wet and/or dusty cartons will be rejected. A quality report must accompany every consignment.

Inspection: Regardless of any accompanying quality report or certificate, the normal entrance control checks will be carried out on every consignment as outlined elsewhere in the company's QC Procedures Handbook. Any detected defects that are over the tolerance levels specified will be rejected if a repeat sampling confirms a higher level of defects above the tolerance levels as indicated below.

Classification of Defects

Critical or Class A Defects: These are faults that make the cartons unusable or, if used, may pose a threat to the survival of the product. Examples are tears, holes, glue failure, odour, wetness, wrong grain direction (if carton has to be run on automatic machine), asymmetric layout, etc.

Major or Class B Defects: These are faults that result in marginal functionality or low-quality appearance. Examples are low grammage, small deviation in dimensions, colour variation, smudged printing, poor crease quality, poor stability to UV light, etc.

Minor or Class C Defects: These are minor faults affecting mostly appearance. Examples are slight off-colour, rough printed surface or low gloss, improper packing for delivery, etc.

Allowable Defect Levels:

Class A or Critical	0.5%
Class B or Major	2.0%
Class C or Minor	5.0%

Prepared by: --- Date:

Checked by: --- Date:

Approved by: --- Date:

Question 103 – Write a specification for a 3-piece tinplate can for a 450 g milk powder

Company Name: XYZ Company Ltd
Packaging Item: 450 g Milk Powder
Ref. number: xxxxxxxxxxx
Date: 12 April 2008

Scope: This specification describes the tinplate can for packing 450 g of milk powder for mass distribution in the local market.

Construction:

    (a) Body: The body shall be made of 0.20 mm Temper 3 tinplate, tin coated to E2.8/E2.8 (2.8 g/m² on each side). Body shall be printed to the approved design (Ref. Artwork, etc.) without internal lacquer. Side seam shall be by welding process. This must be followed by post welding lacquer application to the side weld line both internally and externally with adequate curing.

Some Basic Body Dimensions:

Open Can Height	135.20 ± 0.3 mm
Inside Can Diameter	99.30 ± 0.1 mm
Flange Width	2.50 ± 0.2 mm

Ref. Drawing Number ---

    (b) Fitted End: Both the ring and the lid/cap are stamped from an 0.20 (or 0.19) mm Temper 3 E2.8/E2.8 externally lacquered tinplate. The ring shall have an epoxy enamel lining compound uniformly applied to its grooves followed with adequate curing. The compound must be of food grade and be odour-free. The ring must be fitted with an embossed 0.08 mm aluminium foil diaphragm. Finally the ring and the lid/cap must form a good fit.

See Ref. Drawing Number ---.

Some Key Seam Dimensions/Parameters:

Body Hook Length	2.03 ± 0.15 mm
End Hook Length	2.03 ± 0.15 mm
Seam Thickness	1.20 ± 0.10 mm
Seam Length	2.9 – 3.1 mm (typical)

(c) Loose (Bottom) End: This is stamped from 0.20 mm (or 0.19 mm) Temper 3 E2.8/E2.8 externally lacquered tinplate. A lining compound must be applied to the end with proper curing, just like the ring in (b) above.

Some Key End Dimensions/Parameters:

Outside Curl Diameter	108.7 ± 0.1 mm
Countersink Depth	3.0 ± 0.1 mm
Curl Height	2.0 ± 0.1 mm
Countersink Diameter	99.2 ± 0.1 mm

End profile (Ref. Drawing Number xxxx)

Printing: Printing shall be by offset lithographic process to the approved artwork and agreed colour guides (Ref. yyyy). The printing inks, the enamel coating, and the varnish (if any) must be stable to prolonged exposure to UV light and all rub/scuff resistance tests. The enamel coating colour must be pure and unadulterated white and must not exhibit off-white shade with ageing. The lacquer must be stable to 10% $CUSO_4$ solution. The supplier/printer is responsible for the quality and integrity of all printing inputs, such as inks, lacquer, and enamel coating.

Conversion: Since this is the heart of can making, both the end and the side seams must be leak-free, and be hermetical in integrity. All seam evaluation measurements must comply with the specified/agreed values.

Packing and Delivery: Cans must be delivered with the fitted ends seamed on and the bottom ends packed and delivered separately. Cans must be packed, preferably inverted, on standard four-way entry pallets. Clean and suitable layer boards must be put between adjacent layers of cans. Pallet loads must be supported or wrapped on all sides and the pallets shrink/stretch wrapped and strapped.

Inspection: Regardless of any accompanying quality report or certificate of analysis, every consignment will be subjected to the normal entrance control checks as outlined elsewhere in the company's *QC Procedures Handbook*. Any detected defects, after a repeat sampling and checking that are over the specified tolerance levels indicated below will be rejected.

Classification of Defects
Class A or Critical Defects – These are faults which prevent a can from safely containing and preserving/protecting the product. Examples of such defects are

- Leakage anywhere on the body seam or end seam

- Missing seaming compound from any end or incomplete ring of compound

- Cracks of any size in body or end

- Dented flange at open end, of such degree as to prevent complete double-seam formation

- Contamination of any can interior with rust, or any other foreign matter which is not removable in the can washer

- Missing side seam stripe

- Missing or incomplete or scratched lacquering on inside of body or end, if internal lacquering is required

Class B or Major Defects – These are faults that make a can of border-line functionality or seriously deficient in appearance. Examples are

- Dents over 25.4 mm long.

- Out-of-round at open end, such that closing effectiveness and efficiency may be reduced or impaired.

- Seaming compound below minimum weight.

- Outside enamel missing or incomplete.

- Excess compounds in ends resulting in ends sticking together and consequently causing jams in feeding from stacks.

Class C or Minor Defects – These are faults which adversely affect appearance, but not functionality. Examples are

- Dents less than 25.4 mm long.
- Scratches on exterior surface.

Allowable Tolerance Levels

Class A or Critical Defects	0.5%
Class B or Major Defects	2%
Class C or Minor Defects	4%

Prepared by: ------------------------------------- Date:

Checked by: ------------------------------------- Date:

Approved by: ------------------------------------- Date:

Question 104 – Write a specification for a corrugated case to contain 48 cartons of 200 g of detergent powder each of which has outside dimensions of L118 × W36 × H180 (mm).

Company: XYZ Company Ltd
Packaging Item: Corrugated case for 200 g × 48 Detergent Powder
Ref. Number: xxxxxxxxxxxx
Date: 12 April 2008

Scope: This specification describes the corrugated box for packing 200 g × 48 cartons of detergent powder for mass distribution in the local market.

Box Construction: The box shall be single-wall corrugated RSC (0201) style with 112 gsm 'B' or 'C' flute. The outer and the inner liners (facings) shall be made from long softwood fibres and made by the sulphate pulping process. The resulting kraft liners must be naturally brown with material substances or basis weights as specified below. OFOTB (Outer Flaps Only To Butt).

Note: The use of test liner as inner or outer liner is unacceptable.

Dimensions: Length = 480 mm; Width = 220 mm; Height = 368 mm (inner cartons arranged 6 × 4 × 2 inside the case in an upright position).

Board grade: 200K/112B/150K

All basic dimensions, in mm, shall be interpreted as internal, measured from one score line to the next score line with tolerance of +2 mm. Slot width shall be 6 mm – see drawing No. xxxxxxxxxxxx for detailed construction of RSC (0201) style corrugated box.

Notes:
   (1) Tolerances on basis weight or material substance shall be ± 5%, while tolerances on basic dimensions shall be +2 mm.

   (2) 'C' flute is allowed in place of 'B' flute.

(3) Where the exact grammage specified for a kraft liner is not available, the sum of the grammages for the outer and the inner kraft liners must not be less than the sum specified for the board. For example, for a specified boardgrade of 200K/112B/150K, 225K/112B/125K is allowed, but the higher grade kraft liner must be on the outside.

(4) The use of a higher grammage fluting medium as a compensation for a lower grammage kraft liner is unacceptable.

'B' AND 'C' FLUTE CHARACTERISTICS

	Number of flute per		Flute Height		Take-up or corrective factor
	Meter	Foot	mm	in	
B-flute	150 – 184	50 – 53	2.1 – 2.9	3/32 (0.063)	1.30 – 1.35
C-flute	120 – 145	42 – 45	3.5 – 3.7	9/64 (0.141)	1.40 – 1.45

Packing for Delivery

Corrugated cartons/boxes are to be wrapped up with kraft paper or covered top and bottom with corrugated board, tightly strapped, and delivered in 500s or in 1000s on pallets.

Inspection: Regardless of any accompanying quality report or certificate of analysis, every consignment will be subjected to the normal entrance control checks as outlined elsewhere in the company's QC Procedures Handbook. Any detected defects, after a repeat sampling and checking, which are over and above the specified tolerance levels indicated below will be rejected.

Classification of Defects

Class A or Critical Defects – These are faults that make the boxes unusable or, if used, may pose a threat to the survival of the product. Examples of such defects are

- Dimensions outside the specified limits
- Open manufacturer's joint
- Wet or rain-beaten board or box
- Severe de-lamination
- Boardgrade more than 7.5% below specified grade
- Contaminated boxes
- Tears or holes in the boxes
- Wrong printing

Class B or Major Defects – These are defects which make the box of borderline functionality or seriously deficient in appearance. Examples are

- Glue bond in corrugated board partly incomplete
- Manufacturer's joint not squared-up
- Illegible or incomplete printing
- Box flaps do not fold easily along score lines

Class C or Minor Defects – These are faults which adversely affect the appearance, but not functionality. Examples are

- Weak or slightly off-colour printing
- Stains, scratches, or scuff marks
- Colour variation

ALLOWABLE DEFECT LEVELS

Class A: 0.5%

Class: B: 2.0%

Class C: 5.0%

Prepared by: ---------------------------------- Date:

Checked by: ---------------------------------- Date:

Approved by: ---------------------------------- Date:

Chapter 11 Miscellaneous Topics in Packaging

Question 105 – Describe briefly the importance of each of the following in the packaging of chemicals:

(i) Security

(ii) Convenience

(iii) Cleanliness of the product

(i) Security – Particular attention must be paid to the security of products when considering package style. In the case of non-hazardous products, the aim is to ensure that the product gets to the final destination with minimum damage or loss either through leakage or by pilferage. The package must protect the product against all hazards, be it mechanical or environmental. For instance, weight loss due to moisture loss or any other packaging defects is unacceptable because the nature of the product dictates that precautions should have been taken against such faults at the packaging design stage. Also, adulteration must be prevented by whatever pilfer evident closure is appropriate for the container.

(ii) Convenience – This is a factor that has nothing to do with the quality of the product itself, but with the packaging aspect of it. Let us consider dispensing as an example. Without a proper dispensing device, many products in the market today will be almost useless to the consumer. Examples are aerosol packages, suppositories, eye and ear drops, etc. Also, when it comes to package size, it is an established fact that different package sizes appeal to different audiences. Just as a homemaker doesn't want a gallon can of floor wax, a janitor doesn't want a 16 oz bottle of floor wax, either. Even at the industry level, while it is well known that there is an economy of scale to be gained with big package sizes, in some cases such sizes have been discouraged because of the limitations of handling facilities at the receiver/ user's end.

(iii) Cleanliness of the product – It will be assumed that the product itself is clean at the onset of filling. Then there is the need to ensure that packaging does not become a source of contamination. Most liquids will not store well in many containers without adequate internal lining or lacquering. Steel drums without suitable lacquers or liners will get corroded in contact with moist products. This is not all. The product itself will get discoloured and contaminated in the process. While the emphasis on the external surface of the container is minimal, a situation whereby product spillage could deface the container surface and/or its label should be guarded against.

Question 106 – Write short notes on the following classes of Intermediate Bulk Containers (IBCs) including advantages and disadvantages of each:

(i) Rigid returnable

(ii) Collapsible returnable

(iii) Expendable

(i) Rigid returnable: IBCs have been defined as containers of size greater than the largest conventional package normally used, but smaller than a bulk vehicle. Rigid returnable (IBCs) are used for the carriage of liquids and solids. They may be made from stainless steel, mild steel, coated steel, light alloys, and plastics. These containers have the advantages of low cost when the initial production is spread over their lives and number of trips each container can make over its lifetime. The main disadvantage is the handling, which could create problems to some users who don't have adequate facilities. Also, since these containers are rigid, they occupy the same space in vehicles on the return trip (when empty) as when they contain materials or products.

(ii) Collapsible returnable IBC: These are made from woven fabrics coated with PVC or a natural or synthetic rubber. Those for liquids must have low centres of gravity and often take the form of sausage-shaped tanks strapped to a transport flat. Product discharge is by external atmospheric pressure and hence no air vent is required. Because they are collapsible, they don't take much space in the vehicle during the return trip.

(iii) Expendable IBC: These are valuable containers for overseas export where returnable IBCs are uneconomical. The commonest types are made from corrugated fibreboard. Another type is the union kraft paper. The bodies of these containers are principally octagonal in order to provide the necessary stiffness, and if they do not have pallet bases, they are provided with some form of runner to enable the insertion of forks from a fork truck.

Question 107 – Write a short essay (about 300–400 words) on bulk transport versus conventional package for chemicals.

By bulk transport we mean the delivery or transport of materials in large quantities per trip. There are many materials that cannot be transported other than by bulk transport. A good example is crude petroleum oil from the oil field to a refinery.

Considering packaging and transport costs, bulk transport of liquid materials is cheaper than the conventional package loads. This can only be explained by the economy of scale theory. However, in choosing between bulk transport and conventional package loads, some important factors must be considered. Otherwise, what initially looks an attractive option may turn out to be operationally difficult. Therefore, when the quantities and distances involved are such that either bulk or packaged loads are feasible, we must consider whether facilities for bulk unloading and storage are available at both the seller and the buyer's premises, the chemical and physical nature of the product, the cost of the product, and the use to which the product will be put at its destination.

First we consider the distances. On a per ton basis, it is cheaper to use bulk transport such as tankers for liquid materials than to use the conventional package loads over any distant journey where both methods of transport are feasible.

Next is the availability or non-availability of unloading and other handling (e.g. dosing) facilities for bulk loads. This factor is very important and cannot be overlooked. Let us consider an example of a proposal to receive Na_2SO_4 powder in woven 1-ton bags instead of in conventional 50 kg multi-ply paper sacks. The proposal was based on the assurance that some cost savings would accrue to the user company if the change was accepted. After some preliminary investigations, which included discussion with the production team that normally uses the chemical, and a look at the available dosing facilities, the proposal had to be rejected for lack of adequate handling facilities. If such a decision could be reached in respect of ordinary 1-ton IBC, one can imagine what additional constraints could be expected in respect of bulk loads that are bigger and heavier than IBC loads.

Another factor is the rate of usage or consumption. There are many materials that, once opened, do not last long. If such materials are delivered in bulk containers to a consumer who will use them several times over a long period of time, deterioration may result before the material is exhausted due to frequent exposure to air and moisture. Also relevant is the capital tied down in terms of material inventory costs.

Advantages of bulk transport are mass transport of goods and the attendant economy of scale benefits. Disadvantages are the investment in expensive handling facilities and the fact that small users cannot benefit from it.

Advantages of conventional package loads are easy handling and accessibility to small users. Disadvantages are that packaging material costs are higher per ton of product.

Question 108 – State the three basic considerations needed in the preliminary assessment for packaging a capstan lathe. Outline three or four points under each consideration related to the export of such a lathe from UK to Warsaw in Poland.

The three basic considerations needed in the preliminary assessment for packaging a capstan lathe are

(i) Facts about the capstan lathe itself

(ii) Facts about the journey hazards the lathe is likely to encounter between the end of the manufacturing line and the user's premises

(iii) Facts about the marketing considerations that may be involved

A lathe is a versatile machine tool and is available in a variety of forms designed for low and high production operations.

Under the product consideration, one should consider the fragility of the lathe, its weight, and the possibility of product damage by mechanical and climatic factors. The journey hazards appear to be the one that calls for most attention in transporting the lathe machine from the UK to Warsaw, a journey likely to involve both the sea and land transportation. A lathe is a heavy equipment and will require following the general guidelines for packing such pieces of equipment for transportation. For example the rigid-bed and frame of the lathe should be firmly secured to a strong platform so as to eliminate all possible movement.

The platform should be used for lifting the lathe rather than its package. All delicate parts that are detachable should be dismantled from the lathe and packed carefully, if necessary, with cushioning material. If dismantling is not possible, all overhanging parts should be prevented from sway, swing, and so forth by bracing them in every direction while making sure that all corners are blocked in. The lathe itself should be packed in a wooden container lined with a waterproof material.

The main climatic hazard that may happen to a lathe is corrosion. While packaging can alleviate this problem to some extent by the use of waterproof liner, the application of corrosion preventives to the relevant machine parts prior to packing will go a long way in solving corrosion problems.

Most of the points under marketing consideration are irrelevant as far as the export of a capstan lathe to Warsaw is concerned. For example, such factors as sales appeal, retail marketing, and promotion are not important since a lathe is not an item that can be displayed on the shelf for the purpose of soliciting customers. All that is important here is adequate information on how to open and handle the pack without damage and how to install or assemble it at the user's premises.

Question 109 – Differentiate between a 'clean' and a 'claused' bill of lading.

The bill of lading performs three functions:

    (i) It is evidence of the contract of carriage

    (ii) It is the document of title to the goods

    (iii) It is the receipt of the goods

By signing a bill of lading, the ship owner or his agent agrees that the goods were received in apparent good order and condition. Any such bill of lading is called a 'clean' bill of lading.

On the other hand if the ship owner received the goods in doubtful or unsatisfactory conditions or packed in unapproved packaging materials or one whose integrity is in doubt, the ship owner issues a 'claused' bill of lading to provide him with a technical defence. In short while a 'clean' bill of lading acknowledges receipts of goods in perfect conditions, the 'claused' one makes a provision for registering one's reservation on the state in which the goods were received.

Question 110 – 'Value Analysis is an analytical technique designed to examine all the components of cost and function of an existing product to determine whether overall cost can be reduced whilst retaining all the original functional and quality requirements.'

This definition of Value Analysis assumes that unnecessary cost has been built into the product during design and production stages. Give three reasons as to why such unnecessary cost can arise, illustrating each by an example from your experience.

Three reasons why unnecessary costs might have been built into a product are

(i) Limited knowledge of product behaviour in a given environment

(ii) Technological improvement/market sophistication

(iii) Doubts about quality of distribution and shipping container capability resulting in deliberate over-packaging

On reason (i) above, it is not unusual for designers of packaging, machines, or even pieces of furniture to build unnecessary costs into products when their knowledge of its performance in an environment is not yet fully known or understood. A company was once in a situation of specifying 140 g target or packed weight for a toilet soap for which 120 g was declared on the wrapper. The reason for this is that soaps normally lose substantial moisture in a tropical environment. But after acquiring enough data on the moisture loss trend over a period of time with some statistical analysis of data on tablet weight distribution, it was possible to reduce the target weight down to 130 g, a reduction of 7 per cent in product give-away.

On (ii), there is no doubt about it that technological improvements have resulted in many cost-saving changes all over the world. Over the years, there have been considerable reductions in the weight of machine components without any corresponding reduction in performance and quality. A very glaring example of this is the weight of motor cars of the fifties and sixties compared to what they are today. Another cost reduction factor is the substitution of cheaper and lighter materials for heavy expensive ones in packaging materials, (e.g. plastics in place of metals) caused a reduction in soda (soft drink) can weights over the years.

A shipping container will be used to illustrate (iii). On occasions, while launching a new product, packaging development has had to initially adopt shipping containers that are probably stronger than what is actually needed to do the job. The simple reason is that one may not be sure of what will actually happen to the package in the trade. So the idea is to play it safe a while by initially indulging in over-packaging. But with some experience, the grade of such shipping containers may be gradually brought down to an acceptable level without causing any damage to the product in the process.

Question 111 – Outline the steps in Value Analysis Procedure illustrating your answer by reference to a pack for an electrical appliance consisting of a corrugated case, closed with gum strip and protected with foamed polystyrene fitment and a polythene bag.

The steps/stages in Value Analysis Procedure are

(i) Information – This is essentially an information gathering stage. This involves finding out every detail about the product itself and its packaging. This includes such things as product/packaging specifications and their costs, product and packaging functions including sales appeal, if relevant.

(ii) Speculation – This is the core of value analysis where ideas are put forward to reduce costs without lowering functional performance or quality. Here every member of the team is free to put forward his/her idea, no matter how weak the idea may sound initially. For example, in the electrical appliance under consideration, someone may feel the polythene bag should be discarded. His point will be noted down for further scrutiny later.

(iii) Evaluation – After ideas have been exhausted and noted, the evaluation begins. Each idea is now discussed carefully using cost of function or estimates as a guide. All those who are in position to influence costs, like the suppliers, should be brought in or consulted. Before any idea is dropped altogether, every team member must be satisfied that the idea will not work. All accepted ideas are further assigned to the team members who are knowledgeable in the affected areas for further study, and a progress report on these studies should be submitted.

(iv) Report and recommendation – As each report is done and fully agreed upon by the team, an official recommendation for the adoption of the proposal should be made, preferably on a suitable document. All agreed proposals should spell out in detail the effect of changes and relevant cost implications/savings.

(v) Implementation – This is where the proposals, if and when approved, are put into effect. However, before full implementation, some laboratory tests or field trials are necessary to ensure that the recommendations, with all their envisaged cost savings, do not land the company in a mess due to inaccurate deductions and estimates. There must be some correlation between the laboratory QC tests and the outcome of the field trial; otherwise there may be no justification for going ahead with full implementation. Finally, for Value Analysis projects to succeed, all stakeholders must be fully committed to every aspect of the project.

References

1. Soroka, Walter; *Fundamentals of Packaging Technology*, 3rd edn. (Naperville, Illinois: Publisher, Year).

2. Soroka, Walter; *Fundamentals of Packaging Technology*, UK 2nd edition, published by IOP, UK.

3. Correspondence Course, *The Institute of Packaging*, UK (1972).

4. Roberston, Gordon l; *Food Packaging, Principles and Practice.*

5. Leonard, Edmund A; *Packaging Specifications, Purchasing and Quality Control* (4th edition).

6. Leonard, Edmund A.; *Packaging Economics* (1980).

7. Hirsch, Arthur; *Food Flexible Packaging – Questions and Answers.*

8. *The Stability and Shelf-life of Food*, edited by David Kilcast and Persis Subramaniam.

9. Stewart, Bill; *Package Design Strategy.*

10. *Packaging: The Facts*, IOP, UK.

11. Paine, F.A; *The Packaging Media* (1977).

12. Sacharow, Stanley and Brody, Aaron L.; *Packaging: Introduction* (1987).

13. Sacharow, Stanley; *A Packaging Primer* (1978).

14. Sacharow, Stanley; *A Guide to Packaging Machinery* (1980).

15. Stewart, Bill; *Packaging as an Effective Marketing Tool.*

16. Selke, Susan E.M.; *Understanding Plastics Packaging Technology.*

17. Guss, Leonard M.; *Packaging is Marketing* (1981).

18. *Handbook of Procurement of Packaging;* PRODEC Helsinki (1989).

19. *Fundamentals of Packaging*, edited by F.A. Paine.

20. *The Wiley Encyclopedia of Packaging Technology;* 2rd edition, edited by Aaron L. Brody and Kenneth S. Marsh.

21. *IOP Membership Qualifying Examination – Past Question Papers* (1982–1989).

22. *The Packaging Encyclopedia*, 1983.

23. *The Packaging Encyclopedia*, 1985.

24. *The Packaging Encyclopedia*, 1989.

About the Authors

 Sola Somade is an industrial/analytical chemist, having received a BSc degree in industrial chemistry from the City University, London, and an MSc in analytical chemistry from the Imperial College, University of London. She worked for Unilever Nigeria Plc for twenty years, where she held several key positions in product development, quality assurance, packaging development, and packaging buying. She was a diploma member of the Institute of Packaging, UK, and she is now a Fellow of the Institute. She is the co-founder of Superior Packaging Consultants Ltd – a firm of packaging consultants and trainers based in Lagos, Nigeria.

 Tunji Adegboye holds a BSc degree in chemical engineering from the University of New Brunswick, Fredericton, Canada, and an MS degree in operations research and statistics from the Rensselaer Polytechnic Institute, Troy, New York, USA. He worked for both Unilever Nigeria Plc and Cadbury Nigeria Plc for a total of twenty-seven years where he held such key positions as quality assurance manager, packaging development manager, and technical manager among others. He was a diploma member of the Institute of Packaging, UK, and is now a Fellow of the Institute. He is the co-founder of Superior Packaging Consultants Ltd – a firm of packaging consultants and trainers based in Lagos, Nigeria.

Index

material for bread, 100–101
paper labels, types of, 7–8
in pharmaceutical industry,
3–4
for pharmaceutical products,
105–106
as silent salesman, 8–10
specification, 119–121
woods for, 43
packaging specification
for corrugated case, 128–131
for folding carton, 121–123
for tinplate, 124–127
significance of, 119–121
package testing, 107
conditions, 108
field trials, 108–109
packaging units
composite containers,
characteristics of, 13–14
moulded pulp containers,
14–16
palletization, 69–70
Patra rub, 116
permanent gases, 35–36
pharmaceutical industry,
packaging in, 3–4
plastic film production, 23–24
plasticisers, 19
plastics
additives, 19
kinds of, 17–18
polypropylene copolymer, 19
polyvinyl acetate (PVA), 68
polyvinyl chloride (PVC), 18
polyvinylidene chloride (PVDC),
19

potential energy, 42
press and blow process, 59
pressurised dispenser. *See* aerosol
printing
in paperboard carton, 11
processes of, 5
protection, 106
pulps
differences in, 45
long-fibred, 46–47
puncture test, 112–113

Q

quality control test, 107
quality report
for inspection, 126, 129
quality requirements, 103

R

reports, 141. *See also* quality
report
rationalization, 116
resilient materials, 63
rosin, 44
rubber, 63
rub testing, 115–116

S

safety, 106
sealing. *See* closures
setting time, 48
shelf-life, 98–99
shock absorption. *See* cushion
shock damage, preventing, 37–38
shock distribution, 38

www.ingramcontent.com/pod-product-compliance
Lightning Source LLC
Chambersburg PA
CBHW030741180526
45163CB00003B/881